From Finite to Infinite Dimensional Dynamical Systems

NATO Science Series

A Series presenting the results of scientific meetings supported under the NATO Science Programme.

The Series is published by IOS Press, Amsterdam, and Kluwer Academic Publishers in conjunction with the NATO Scientific Affairs Division

Sub-Series

I. Life and Behavioural Sciences	IOS Press
II. Mathematics, Physics and Chemistry	Kluwer Academic Publishers
III. Computer and Systems Science	IOS Press
IV. Earth and Environmental Sciences	Kluwer Academic Publishers

The NATO Science Series continues the series of books published formerly as the NATO ASI Series.

The NATO Science Programme offers support for collaboration in civil science between scientists of countries of the Euro-Atlantic Partnership Council. The types of scientific meeting generally supported are "Advanced Study Institutes" and "Advanced Research Workshops", and the NATO Science Series collects together the results of these meetings. The meetings are co-organized bij scientists from NATO countries and scientists from NATO's Partner countries – countries of the CIS and Central and Eastern Europe.

Advanced Study Institutes are high-level tutorial courses offering in-depth study of latest advances in a field.
Advanced Research Workshops are expert meetings aimed at critical assessment of a field, and identification of directions for future action.

As a consequence of the restructuring of the NATO Science Programme in 1999, the NATO Science Series was re-organized to the four sub-series noted above. Please consult the following web sites for information on previous volumes published in the Series.

http://www.nato.int/science
http://www.wkap.nl
http://www.iospress.nl
http://www.wtv-books.de/nato-pco.htm

Series II: Mathematics, Physics and Chemistry – Vol. 19

From Finite to Infinite Dimensional Dynamical Systems

edited by

J. C. Robinson

University of Warwick,
Coventry, U.K.

and

P. A. Glendinning

UMIST,
Manchester, U.K.

Kluwer Academic Publishers

Dordrecht / Boston / London

Published in cooperation with NATO Scientific Affairs Division

Proceedings of the NATO Advanced Study Institute on
From Finite to Infinite Dimensional Dynamical Systems
Cambridge, United Kingdom
21 August–1 September 1995

A C.I.P. Catalogue record for this book is available from the Library of Congress.

ISBN 0-7923-6975-0

Published by Kluwer Academic Publishers,
P.O. Box 17, 3300 AA Dordrecht, The Netherlands.

Sold and distributed in North, Central and South America
by Kluwer Academic Publishers,
101 Philip Drive, Norwell, MA 02061, U.S.A.

In all other countries, sold and distributed
by Kluwer Academic Publishers,
P.O. Box 322, 3300 AH Dordrecht, The Netherlands.

Printed on acid-free paper

Printed in the Netherlands.

NATO ASI: August 21 - September 1, 1995

Title: From finite to infinite dimensional dynamical systems

ORGANIZING COMMITTEE

P. Coullet (Nice, France)
P. Cvitanovic (Niels Bohr Institute, Denmark)
P. Glendinning (Cambridge, U.K.) *Director*
J. Hale (Georgia Tech, USA)

LECTURERS

J. Ball (Heriot-Watt University, U.K., *now Oxford, U.K.*)
L. Bunimovich (Georgia Tech, USA)
J. Carr (Heriot-Watt University, U.K.)
P. Collet (Ecole Polytechnique, Palaiseau, France)
P. Constantin (University of Chicago, USA)
P. Coullet (Nice, France)
P. Cvitanovic (Niels Bohr Institute, Denmark)
P. Glendinning (Cambridge, U.K., *now UMIST, U.K.*)
J. Hale (Georgia Tech, USA)
P. Holmes (Princeton, USA)
I. Procaccia (Weizmann Institute, Israel)
D.A. Rand (Warwick, U.K.)

GUEST LECTURERS

C. Doering (Los Alamos, USA)
J.D. Gibbon (Imperial College, U.K.)
R.S. MacKay (Warwick, U.K.)

SCIENTIFIC PROGRAMME

21 August - Monday

10.45 *Procaccia:* Scaling phenomena in systems far from equilibrium I

11.45 *Hale:* Analysis and partial differential equations I

14.30 *Glendinning:* Bifurcations in finite dimensions I

16.30 *Bunimovich:* Lattice dynamical systems I

22 August - Tuesday

09.30 *Procaccia:* Scaling phenomena in systems far from equilibrium II

11.15 *Hale:* Analysis and partial differential equations II

14.30 *Guest lecturer: Doering:* Convection, stability and turbulence

16.30 *Bunimovich:* Lattice dynamical systems II

23 August - Wednesday

09.30 *Glendinning:* Bifurcations in finite dimensions II

11.15 *Carr:* Infinite dimensional systems and metastability I

14.30 *Coullet:* Qualitative theory of symmetry breaking in extended systems: patterns and topological singularities I

16.30 *Bunimovich:* Lattice dynamical systems III

24 August - Thursday

09.30 *Procaccia:* Scaling phenomena in systems far from equilibrium III

11.15 *Hale:* Analysis and partial differential equations III

14.30 *Coullet:* Qualitative theory of symmetry breaking in extended systems: patterns and topological singularities II

16.30 *Cvitanovic:* Transfer operator methods in dynamical systems I

25 August - Friday

09.30 *Cvitanovic:* Transfer operator methods in dynamical systems II

11.15 *Carr:* Infinite dimensional systems and metastability II

14.30 *Coullet:* Qualitative theory of symmetry breaking in extended systems: patterns and topological singularities III

16.30 Contributed papers

28 August - Monday

09.30 *Holmes:* Dynamics of coherent structures I

11.15 *Constantin:* Mathematical fluid dynamics I

14.30 *Rand:* Spatial dynamics in ecology, evolution, epidemiology and immunology I

16.30 Contributed papers

29 August - Tuesday

09.30 *Holmes:* Dynamics of coherent structures II

11.15 *Constantin* Mathematical fluid dynamics II

14.30 *Rand* Spatial dynamics in ecology, evolution, epidemiology and immunology II

16.30 *Guest lecturer: MacKay* Dynamics of networks

30 August - Wednesday

09.30 *Collet:* Unbounded domains, spectral problems and bifurcations I

11.15 *Constantin:* Mathematical fluid dynamics III

14.30 *Ball:* Dynamics and variational principles I

16.30 Poster discussion *(Posters will be displayed from day 6 to 8)*

31 August - Thursday

09.30 *Holmes:* Dynamics of coherent structures III

11.15 *Collet* Unbounded domains, spectral problems and bifurcations II

14.30 *Ball* Dynamics and variational principles II

16.30 *Guest lecturer: Gibbon* Regularity and global attractors for the Navier-Stokes equation

1 September - Friday

09.30 *Collet* Unbounded domains, spectral problems and bifurcations III

11.15 *Ball* Dynamics and variational principles III

14.30 *Rand* Spatial dynamics in ecology, evolution, epidemiology and immunology III

16.30 Farewell discussion

CONTENTS

PREFACE

The lectures in this volume were given as part of a NATO Advanced Study Institute *From finite to infinite dimensional dynamical systems* held at the Isaac Newton Institute for Mathematical Sciences, Cambridge, U.K., between 21 August and 1 September 1995. The NATO ASI brought together over sixty young scientists and mathematicians from all over the world to listen to expository talks from leading experts in the field, and also gave them the opportunity to present some of their own results as posters or as lectures. We hope that what they learned both formally and in informal conversations, and the friendships made during their stay in Cambridge, will remain important to them for many years to come.

The NATO ASI formed part of a larger six month project at the Newton Institute on the same topic, and many of the participants were able to stay on in Cambridge to attend a Euro-Conference and both hear and contribute to the research in progress which was reported. We are grateful to all the bodies which helped fund these enterprises; principally, of course, NATO, but the European Union, CNRS (France) and the London Mathematical Society all made contributions which allowed participants to stay on for the research conference and the Leverhume Trust funded the visits of participants from the former Soviet Union. The involvement of the Newton Institute project and an EU Human Capital and Mobility Network (*Nonlinear Phenomena and Complex Systems*) also made it possible to have 'guest lecturers' during the ASI.

The smooth running of an Advanced Study Institute depends on the work of people behind the scenes. The staff at the Newton Institute were responsible for much of the administrative work and were unfailingly helpful and cheerful despite the large number of participants and lecturers involved. Without their help it would have been impossible to mount a NATO ASI on this scale.

Ultimately, the scientific success of a NATO ASI depends on the lecturers and the participants. Every lecturer made great efforts to be comprehensible, interesting and enthusiastic. We are grateful for the time and energy they put in to this project. To the participants all we can say is thank you and we hope to see you again. The participants generated an atmosphere of serious scientific enquiry and also a co-operative and friendly ambiance which greatly enhanced the experience for all of those involved.

Finally, we would like to say a personal thank you to Colin Sparrow. The existence of the Newton Institute project, and hence the impetus for the application to NATO in the first place, was due to his vision and hard work. His guiding hand and wise words made the whole thing possible.

The articles collected together in this volume treat fundamental properties of partial differential equations or of modelling processes which involve spatial dynamics. Although they were written five years ago we believe that they provide an excellent introduction to the mathematics of problems related to spatial complexity.

James Robinson & Paul Glendinning
Warwick and Manchester
January 2001

LIST OF CONTRIBUTORS

L A BUNIMOVICH
Center for Dynamical Systems
& Nonlinear Studies
Georgia Institute of Technology
Atlanta, GA 30032-0190
USA

P COLLET
Centre de Physique Theoretique
Ecole Polytechnique
91128 Palaiseau Cedex
FRANCE

P CONSTANTIN
Department of Mathematics
University of Chicago
5734 South University Ave
Chicago, IL 60637
USA

P A GLENDINNING
Department of Mathematics
UMIST
P O Box 88
Manchester M60 1QD
UK

J K HALE
School of Mathematics
Georgia Institute of Technology
Atlanta, GA 30032-0190
USA

P J HOLMES
Department of Mechanical and
Aerospace Engineering
Princeton University
Princeton, NJ 08544-5263
USA

M J KEELING
Zoology Department
Cambridge University
Downing Street
Cambridge CB2 3EJ
UK

J C MATTINGLY
Department of Mathematics
Building 380
Stanford University
Stanford, CA 94305-2125
USA

D A RAND
Mathematics Institute
University of Warwick
Coventry CV4 7AL
UK

J C ROBINSON
Mathematics Institute
University of Warwick
Coventry CV4 7AL
UK

R W WITTENBERG
Department of Mathematics
University of Michigan
2074 East Hall
525 East University Avenue
Ann Arbour, MI 48109-1109
USA

INTRODUCTION

J.C. ROBINSON
Mathematics Institute
University of Warwick
Coventry CV4 7AL.
U.K.
email: jcr@maths.warwick.ac.uk

AND

P.A. GLENDINNING
Department of Mathematics
UMIST
Manchester, M60 1QD.
U.K.
email: p.a.glendinning@umist.ac.uk

The theory of finite-dimensional dynamical systems – the qualitative study of the solutions of ordinary differential equations and iterated maps – was essentially begun at the end of the last century by Poincaré. With the advent of increased computing power the subject received a considerable boost, and has developed dramatically over the last thirty years. Since the influential papers of Lorenz (1963) and Smale (1967), there has been much progress, and, partly because of its visual appeal, the subject has gained some public prominence under the media-friendly chaos banner.

Indeed, the intuition provided by the visual element of computer simulations has played a part in setting somewhat artificial boundaries to the problems studied – the ease of visualising the three-dimensional Lorenz system (for example) has proved a powerful stimulus for its study. Of course, such low-dimensional systems are also mathematically more tractable, although despite a fairly complete understanding of one-dimensional maps (see de Melo & van Strien, 1993, for a comprehensive survey), the two-dimensional case is still understood only in some special cases (e.g. Mora & Viana, 1993).

It is perhaps no surprise, then, that the application of such ideas to the infinite-dimensional problems that arise in the study of partial dif-

J. C. Robinson and P. A. Glendinning (eds.),
From Finite to Infinite Dimensional Dynamical Systems, 1–3.

ferential equations has had to wait until relatively recently. Within the last twenty years, however, many of the qualitative ideas from the theory of finite-dimensional differential equations have been generalised to the infinite-dimensional case. Much of this work was first done in the context of delay differential equations by Hale and co-workers (e.g. Hale, 1971: Hale *et al.*, 1974), then generalised to the PDE case (e.g. Henry, 1981; Hale, 1988); great advances have also been made in recent years by those studying the PDE models of fluid dynamics (Constantin *et al.*, 1985; Constantin & Foias, 1988; Temam, 1979, 1988a & b; Ladyzhenskaya, 1991).

The papers in this volume represent a variety of different approaches, from the study of bifurcations in low-dimensional systems, through biological systems with extended spatial structure, to infinite-dimensional dynamics and various approaches to their complexity.

The first two papers treat spatially extended systems as an intermediate step between finite-dimensional systems and partial differential equations. Matthew Keeling and David Rand apply many different tools from the theory of dynamical systems to investigate assorted biological problems, including the evolution of virulence and the advantages of sexual reproduction. Leonid Bunimovich treats similar systems from a more abstract point of view, introducing the idea of a lattice dynamical system, a yet more general version of the coupled map lattice model, which has been much studied and is used throughout the paper by Keeling & Rand.

Jack Hale provides a general overview of the theory of infinite dimensional dynamics, and in particular discusses various properties of the global attractor. This is a compact invariant set which attracts all trajectories of the equation, and in many cases it can also be shown to be finite-dimensional. Studying the dynamics on this set is equivalent to concentrating on the asymptotic behaviour of the system, a ubiquitous idea in dynamical systems.

Pierre Collet's approach is an analytic one, and he addresses the difficulties which arise when the model equation is posed on the whole line rather than only on an interval. To illustrate the powerful techniques which have been developed to deal with this situation, he uses various example equations, including Burgers' equation and the Ginzburg-Landau equation.

In contrast to this, the analysis contained in the lectures by Peter Constantin, which study the Euler and Navier-Stokes equations of fluid dynamics, is highly constrained by the nature of the equations themselves. Indeed, existence and uniqueness results for the three-dimensional Navier-Stokes equations, and a non-existence result for the Euler equations, are two of the grails of the subject (see Doering & Gibbon, 1995, for a nice discussion of these problems). Some very subtle analysis is needed even to begin to approach these problems.

The final paper in this volume, by Philip Holmes, J. Mattingly, and Ralf Wittenberg, addresses a natural problem that confronts anyone who wishes to study the physical systems which are modelled by partial differential equations; namely, how to find an analytically (or computationally) tractable model which can be expected to retain the essential features of the original system. With a mixture of physical intuition, computation, and dynamical systems-inspired methods, they discuss the turbulent boundary layer near a wall using a low-dimensional system of ordinary differential equations.

We hope that the selection of papers in this volume will not only give a useful introduction to the application of dynamical systems ideas to the study of infinite-dimensional equations, but also serve to highlight the broad range of topics in which such ideas can provide both the questions and, in some cases, the answers, which more traditional approaches would not have suggested.

References

Constantin, P. & Foias, C. (1988) *Navier-Stokes equations*, University of Chicago Press.

Constantin, P., Foias, C., & Temam, R. (1985) Attractors representing turbulents flows, *Mem. Amer. Math. Soc.* **53**.

Doering, C.R. & Gibbon, J.D. (1995) *Applied analysis of the Navier-Stokes equations*, Cambridge University Press.

Hale, J.K. (1971) *Functional differential equations*, Springer Appl. Math. Sci. Volume 3: see also Hale, J.K. & Verduyn Lunel, S.M. (1993) *Introduction to functional differential equations*, Springer Appl. Math. Sci. Volume 99.

Hale, J.K. (1988) *Asymptotic behavior of dissipative systems (Mathematical Surveys and Monographs 25)*, American Mathematical Society, Providence, RI.

Hale, J.K., Maghalhães, L., & Oliva, W. (1974) *An introduction to infinite dimensional dynamical systems – Geometric Theory*, Springer Verlag.

Henry, D. (1981) *Geometric theory of semilinear parabolic equations*, Lecture Notes in Math., Vol. 840, Springer-Verlag, New York.

Ladyzhenskaya, O. (1991) *Attractors for semigroups and evolution equations*, Cambridge University Press.

Lorenz, E. (1963) Deterministic nonperiodic flow, *J. Atmos. Sci.* **20**, pp. 448–464.

de Melo, W. & van Strien, S. (1993) *One-dimensional dynamics*. Springer Verlag, London.

Mora, L. & Viana, M. (1993) Abundance of strange attractors, *Acta Math.* **171**, pp. 1–71.

Smale, S. (1967) Differentiable dynamical systems, *Bull. Amer. Math. Soc.* **73**, pp. 747–817.

Temam, R. (1979) *Navier-Stokes equations, theory and numerical analysis*, North-Holland, Amsterdam.

Temam, R. (1988a) *Infinite dimensional dynamical systems in mechanics and physics*, AMS 68, Springer, London. (Second edition: 1997).

Temam, R. (1988b) *Navier-Stokes equations and nonlinear functional analysis*, SIAM, Philadelphia.

SPATIAL CORRELATIONS AND LOCAL FLUCTUATIONS IN HOST-PARASITE MODELS

M.J. KEELING
Zoology Department
University of Cambridge
Downing Strret
Cambridge CB2 3EJ
U.K.

AND

D.A. RAND
Mathematics Institute
University of Warwick
Coventry CV4 7AL
U.K.

1. Introduction

The interaction between humans and infectious diseases is one of the most important and best understood areas of ecology. The simple mathematical models which describe the ideal progression of a disease through a well mixed population have been the subject of vast amounts of scrutiny in recent years (Anderson & May, 1992). These models give an accurate portrait of the spread of an infection when there is a large population with homogeneous mixing and little structure or segregation. However, despite our fairly clear understanding of the dynamics produced by these simple epidemiological systems many problems arise when the underlying assumptions no longer hold: unfortunately this is usually the case.

In this chapter we shall study the effects of small local populations, spatial distribution and limited local iterations. Although these complications will be added to only the most rudimentary of models we shall see that surprising results as well as more realistic predictions can emerge. Most of the results can be attributed to two distinct phenomena, the existence of strong local correlations and the large stochastic fluctuations that are seen

J. C. Robinson and P. A. Glendinning (eds.),
From Finite to Infinite Dimensional Dynamical Systems, 5–57.

at the scale of local interactions. Understanding and modelling the effects of these two factors will be the main focus of this discussion and various simple models will be introduced to highlight the typical effects.

This work is divided into five sections; section 2 introduces the framework of a simple spatial model and discusses some of the problems that are encountered. Section 3 concentrates on this model of a very simple host pathogen system and compares the dynamics and evolution of the spatial system to those of the standard differential equations. Section 4 extends this model in several directions but looks in particular at the maintenance of sexual reproduction due to parasitism. Finally, section 5 looks at two methods of simulating the behaviour of the discrete spatial model by more conventional differential systems.

2. The basic model and its associated problems

In our models we wish to include effects due to space, the discreteness of populations, and stochastic fluctuations; *Probabilistic cellular automata* (or *Interacting particle systems*) are therefore a natural medium. Probabilistic cellular automata (PCA) are defined by the following three characteristics;
1) A lattice of distinct sites (or cells), most often in a two dimensional lattice arrangement.
2) Each site is in one of a finite set of states.
3) The local neighbourhood of a site determines the probabilistic behaviour of that site at the next iteration.
PCA and deterministic cellular automata have been used to model a variety of ecological and epidemiological systems, giving very good approximations when one of the components is sessile or slow moving (Mollison, 1977; Ermentrout & Edelstein-Keshet, 1992; Durrett & Levin, 1994a).

The simplest version of our model will now be described to help clarify the type of system we are dealing with. (These rules will be expressed mathematically in §3.1.2, and further refinements will be introduced later as they are needed.) The system consists of a two-dimensional square lattice of sites together with periodic boundary conditions, so that all activity takes place on a torus. This boundary condition is merely for simplicity and has little effect on the dynamics of the system. Each site is in one of three basic states, E, H, or P corresponding to *empty*, occupied by a healthy *host* and occupied by a *parasitised* host respectively; for convenience these sites will often be referred to as Empty, Host and Parasite. Each site is influenced by its nearest four neighbours (North, South, East and West, termed the Von Neumann type neighbourhood). A healthy host can grow into an empty site in its neighbourhood with probability g (infected hosts cannot grow) and can be infected by a parasitised host in its local neighbourhood with probability of transmission T. Initially the parasite is assumed to be fatal and so sites occupied by parasitised hosts become empty at the next iteration as the virulence $V = 1$. The entire lattice is updated synchronously; although it can be argued (Brown, 1993) that the computationally slower asynchronous updating is more natural, no qualitative difference between the two methods has been observed for this model or any of the extensions developed later.

PCA models have several distinct advantages over their deterministic differential counterparts. The rules are in general far more natural as they are specified at the level of the individual, and so can be formed from behavioural observations, allowing the large scale global dynamics to emerge. The PCA as defined above automatically include both space and a carrying capacity of the hosts, limited local interactions together with stochastic fluctuations and a population composed of individuals. Any of these components can affect the dynamics as a profusion of spatial models demonstrate (Barkham & Hance, 1982; Durrett & Levin 1994b; Wilson *et al.*, 1993; Mollison *et al.*, 1993).

The main distinction between spatial and homogeneous mean-field models is the presence of local as opposed to global interactions. Both these behaviours are seen to some degree in the natural world, so often a gradual scale between the two is useful, and this is studied in section 4. It has already been shown (Anderson & May, 1992) that spatial correlations can have a profound effect on the dynamics of an epidemic, and these PCA models demonstrate a natural way in which these correlations can arise. The introduction of space and limited range interactions often leads to the formation of refuges from parasitism for some of the hosts. For example, if the parasites can only spread a very short distance then any isolated group of hosts will be protected. The biological implications of refuges are widely described (Begon *et al.*, 1990); they offer reservoirs from which individuals can disperse into the environment, damping down oscillations and helping to stabilise the dynamics. This stabilising behaviour is exemplified by the spatial Nicholson-Bailey model (Hassel *et al.*, 1991), where the divergent oscillations of the non-spatial system are replaced by more confined dynamics. Spatial models demonstrate simply the existence of ephemeral refuges due to the spatial structure of the populations and the local interactions. These refuges are an example of self-induced spatial heterogeneity: from a homogeneous distribution with no correlations between local sites a heterogeneous spatial pattern emerges due to amplification of stochastic noise. These spatial patterns and the associated local correlations can clearly be seen in the snap-shots of the various systems (Figures 1, 2, and 15).

The consideration of individuals in modelling is most important when dealing with low densities or low numbers of organisms, such as the number within a local neighbourhood. Most models simply deal with densities and make no distinction between small and large populations. Small populations are highly susceptible to probabilistic behaviour and for example there are many documented cases (Pimm, 1991) of extinctions of small populations. In general it is clear that the chance of the population becoming extinct decreases exponentially as the initial size of the population increases. Another implication of a discrete population is that each site experiences an integer number of individuals, and hence its local environment may be forced to lie far from the mean as it cannot include fractional individuals. This integer based neighbourhood will further enhance the effects of stochastic fluctuations which will be relatively large at the local scale. Attempts to simulate stochastic fluctuations and local correlations are dealt with in section 5 and both prove to be necessary in explaining the dynamics of the simple spatial host-parasite system.

Although the use of probabilistic cellular automata brings the advantages of including the realistic features of space, stochasticity and discrete individuals, it also introduces many of the problems associated with real ecosystems.

The results are stochastic, so multiple simulations are always necessary; this phenomenon becomes very obvious when dealing with invasions. In standard deterministic systems typical invasions are either always successful or always fail. With PCAs this is no longer true, as any invasion simply has a probability of success which may be highly dependent on both the number and distribution of the invading individuals. This causes problems with the definition of selective pressure (see Section 3) although the behaviour is far more like a real system. Stochasticity also affects the standard notions of attractors and fixed points; what we usually observe is a low dimensional attractor together with high dimensional noise which complicates many problems. For example, there is still a great amount of debate over the detection of chaotic signals in noisy data or even whether chaotic dynamics exist in real ecosystems (Pool, 1989; Sugihara *et al.*, 1990). In most of this work we obtain fixed point attractors for the global average together with coloured noise. We will term a stochastic time series *statistically stationary* if it has a time independent mean and the fluctuations are small relative to the number of individuals present and the rate of change of the sites.

The stochastic fluctuations are also important when considering the size of the lattice and therefore the sampling size. This issue is addressed more fully in Keeling *et al.* (1997a) but the intuition is as follows. If a given system is observed at too small a scale then there are relatively large fluctuations due to random events; at the other extreme when the system is viewed at too large a scale then any interesting spatial dynamics (such as limit cycles or chaotic attractors) are averaged out and we just observe a global mean. The central limit theorem predicts that as the number of sites in the lattice (N) increases then the variance scales like N^{-1}. We term the *coherence length* the size of lattice at which this scaling behaviour begins; this is the size where the underlying dynamics are maximal compared to the noise and is the size at which all the results are quoted.

We shall now consider the result for the simple Host-Parasite model described above (a more in depth analysis can be found in Rand *et al.*, 1995) paying particular attention to the effects of spatial correlations and large scale fluctuations. This investigation will also illustrate many of the problems and phenomena found in stochastic spatial models and will act as a stepping stone to more complex systems addressed in the following sections.

3. Host-Parasite models (Individuals and Spatial Distributions)

The growth rate of the host into empty cells (g) will be fixed at 0.05 which is slow enough to be realistically less than the transmission rate (T) and yet fast enough so that simulations are feasible. The virulence (V) is also fixed (at 1) so that attention can be focused upon changes in behaviour with T. The behaviour with respect to changes of the virulence of the pathogen is discussed in Rand *et al.* (1995), but the same qualitative phenomena hold. The *coherence length scale* of the system was found to be between 80 and 90 cells. This agrees well with observations, in that the parasite quickly dies out if a lattice size of 70 or less is used, but usually survives on a lattice of 100 cells.

3.1. GLOBAL DYNAMICS OF THE PROBABILISTIC CELLULAR AUTOMATA

When running the cellular automaton it is quickly noticed that the exact spatial state of the system is far too sensitive to stochastic events to allow any form of precise prediction. However, once the initial transient behaviour has disappeared (after 100 or so iterations) the proportions of hosts, parasites and empty sites do not seem to vary by much. Also, the general appearance of the spatial structures seems uniform, so that the system is statistically stationary. All this is confirmed when examining the numbers of each type of site (Empty, Host or Parasite) and the proportion of neighbourhood types, which remain fairly constant, only affected by stochastic fluctuations. (The neighbourhood types refers to the type of site and the number of each type that surround it. e.g. a central host surrounded by 1 parasite, 1 empty cell and 2 other hosts.) So not only are the numbers constant, but so are the local correlations and hence the interactions as these are specified by the local neighbourhoods.

3.1.1. *Existence of a Critical Transmissibility* T_c

This section contrasts the behaviour of the homogeneous mean-field system and of the cellular automaton and demonstrates the existence of a critical transmissibility T_c above which the spatially distributed parasite cannot survive. In the homogeneous model, each cell is affected equally by every other cell, irrespective of its location; the homogeneous model also assumes the space is infinite so the parasite can survive down to unrealistically low levels and still recover to a reasonable population size. With the cellular automaton this is no longer true and a new phenomenon of critical transmissibility emerges due to local correlations. Each site in the lattice is only influenced by its nearest neighbours, so a parasite with too high transmissibility can infect the surrounding hosts so quickly that the population has no time to recover and so in subsequent iterations there are too few healthy hosts next to parasites to sustain the parasite population. The cellular automaton being a discrete stochastic system also enhances this effect as population levels near (number of sites)$^{-1}$ usually die out.

3.1.2. *Homogeneous Mean-Field Equations*

By converting the basic set of rules (given in Section 2) into a coupled difference equation we obtain:

$$\begin{array}{lrcl} \text{Hosts} & \mathcal{H}' & = & \mathcal{H}(1+4g(1-\mathcal{H}-\mathcal{P})-4T\mathcal{P}) \\ \text{Parasites} & \mathcal{P}' & = & \mathcal{P}(1+4T\mathcal{H}-V). \end{array} \tag{1}$$

When $T \geq V/4$ the non-trivial stationary point is found to be:

$$\begin{array}{rcl} \mathcal{H} & = & V/4T \\ \mathcal{P} & = & g(4T-V)/4T(g+T) \end{array} \tag{2}$$

The stationary point is stable for all small values of g compared to T and there is no maximum value of T for which these solutions exist, so in this model $T_c = 1$.

3.1.3. *Behaviour of the Cellular Automaton*

For the cellular automaton, with $g = 0.05$ and $V = 1$, the parasite only survives when T is approximately greater than 0.46, compared to 0.25 in the homogeneous mean-field case. It was expected that this lower bound should have increased (to be at least $\geq \frac{1}{3}$), as an infected host can be surrounded by at most three healthy hosts because it must have been infected by one of its neighbours which would subsequently have died. When running the model for values larger than 0.46, standard behaviour and patterns soon emerge, although the precise dynamics are unpredictable. Fronts of parasites move through the hosts leaving only blank cells in their wake, which are slowly colonised (See Figure 1). For larger transmissibility the fronts are more complete, so the amount of blank cells and the time for recolonisation by a host are increased (Figure 2) until at large values of T the parasites die out.

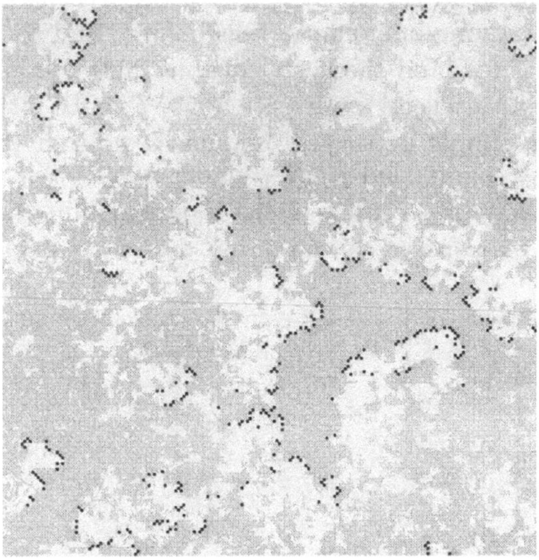

Figure 1. Example of a system with $g = 0.05$ and $T = 0.55$. Empty sites are white, hosts are grey and parasites black.

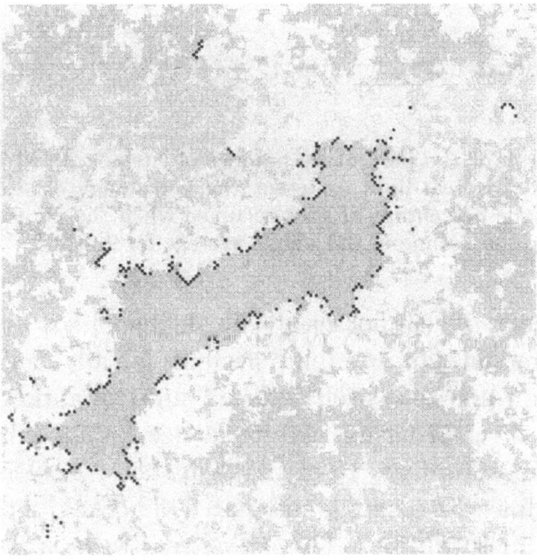

Figure 2. Example of a system with $g = 0.05$ and $T = 0.6$. Notice that there are larger unbroken fronts of parasites and fewer parasites in general.

Increasing T to values greater than approximately 0.6 the parasites start to die out with an ever increasing probability; this brings us to the hypothesis that there exists a critical transmissibility T_c above which the probability of survival is zero. If T is close to 1 then the parasites spread as a wavefront leaving only empty cells behind, and so on any finite sized lattice the parasites (and the hosts) die out fairly rapidly. Figure 3 shows the expected percentage of parasites against T, both if the system survives and also just a simple average. It should be noticed that the number of parasites in the cellular automaton model is far lower than predicted by the mean-field equations, and once the transmissibility has risen above 0.55 the number of parasites in the surviving systems remains fairly constant.

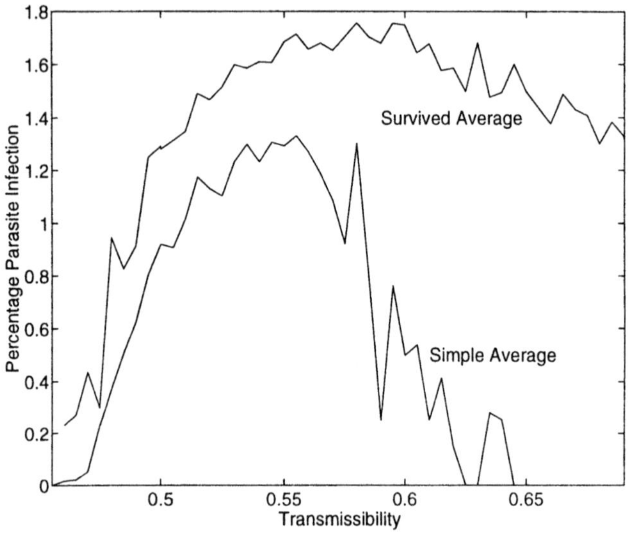

Figure 3. Graph of the percentage number of sites occupied by parasites against transmissibility. The two lines represent the percentage of sites occupied for any simulation and the percentage of sites occupied when the parasite is known to have survived. Notice the decline in the simple average for $T > 0.6$.

Unfortunately, T_c cannot easily be defined because of the stochastic nature of the model. On any finite system, for any finite amount of time, there is always a non-zero probability of survival and of extinction, so the probability of survival on an infinite lattice for an infinitely long period of time must be examined. To accomplish this, consider the probability of surviving starting with a small finite patch of hosts and parasites, and allow the time to tend to infinity. The parasites always spread faster than the hosts, so at some time $t = t_o$ there exists a disc of hosts and parasites which is homogeneous at large scales; this will expand at a near constant velocity v (Van de Bosch *et al.*, 1990). The analysis can therefore be restricted to the survival of the parasite on an $N \times N$ lattice for $\frac{N}{v}$ time steps, where N is large.

It should be noticed that any real-life situation will always have a finite population, corresponding to a finite lattice size, so the parasite cannot be expected to last for ever. If world populations and evolutionary time spans are to be studied then the value of T_c is a close approximation; if we want a parasite to survive on a small island community then T_c will be much lower. A more useful value may be $T_c(t, n, p)$, the transmissibility at which the chance of survival, for t iterations with a population of n hosts, is p. However this depends far too much on the initial conditions to give us a natural value and it would have to be calculated for the specific problem in mind.

3.1.4. *Monotonicity with Respect to Transmissibility*

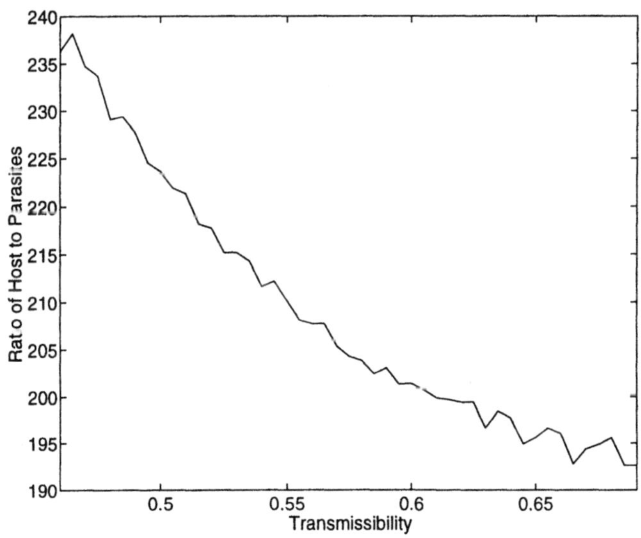

Figure 4. A comparison of the average number of hosts per parasite for a range of transmissibilities. Values for $T > 0.67$ were difficult to obtain due to rapid decline and extinction of the parasite.

Previously the monotonicity of behaviour with respect to the transmissibility was implicitly assumed. Some supporting observational and numerical evidence will now be given. For T close to 1, the parasites spread out as a near perfect circle, leaving only empty sites behind. As T is reduced the spreading circle has more breaks in it and there is now a greater chance of some hosts spreading into the inner empty disc, giving the parasites more cells to colonise and so a greater chance of surviving. Hence for T close to 1 the chance of survival increases as the transmissibility decreases.

For smaller T, a purely mixed state is soon reached where increasing T instantaneously increases the number of parasites hence decreases the number of hosts, decreasing the number of sites to which the parasite can spread which in turn leads to an increased risk of parasite extinction. As can be seen from Figure 4 the number of hosts per parasite is a monotonically decreasing function of the transmissibility, even though the proportion of parasites remains constant (Figure 3), so there must be more empty sites. It is this increase in the proportion of empty sites that acts as a barrier to parasite transmission, and that eventually leads to their extinction.

3.1.5. *Results*

We will now use the method outlined above to find an approximation to the critical transmissibility T_c. For all values of transmissibility the velocity v of the spreading disc of hosts and parasites was found to be close to 0.1, although there was a slight decrease with transmissibility, up to T_c. A transmissibility T was considered to lie above T_c if the parasite failed to survive for the appropriate number of iterations in all of 200 simulations. The approximations to T_c were found for lattice sizes from 80×80 to 300×300 and the results clearly asymptote to around 0.67. This value corresponds well with intuitive observations.

The monotonicity of the system was used to find T_c by means of a 'lion hunt' algorithm; with this method the size of the interval in which we know the critical value resides halves at each step. Although it would only take a few steps of the lion-hunt algorithm to obtain a more accurate approximation (to 3 significant figures for example) it was felt that with only 200 simulations and a maximum lattice of 300×300 being used, greater accuracy could not be justified. Also, as indicated above, the actual value of T_c is fairly irrelevant when it comes to real situations; it is mainly the existence of such a threshold that is important.

3.2. EVOLUTION TO THE CRITICAL TRANSMISSIBILITY

We have observed how this PCA model possesses very different dynamics from its deterministic, homogeneous counterpart on the ecological time scale. To see how the parasite will evolve over longer times, it is necessary to examine the behaviour of the selective pressure. This is a measure of the speed of evolution in a particular direction, and for increasing transmissibility is given by

$$\mathcal{S} = \frac{\text{Initial Growth Rate of a Mutant} * \text{Probability of Mutant Invasion}}{\text{Transmissibility of Mutant} - \text{Transmissibility of Parasite}}$$

at the stationary point.

If $\mathcal{M}$ represents the mutant population, then the homogeneous mean-field equations (cf. Equation (1)),

$$\begin{aligned} \mathcal{H}' &= \mathcal{H} + 4g(1 - \mathcal{H} - \mathcal{P} - \mathcal{M})\mathcal{H} - 4T_P\mathcal{H}\mathcal{P} - 4T_M\mathcal{H}\mathcal{M} \\ \mathcal{P}' &= 4T_P\mathcal{H}\mathcal{P} \\ \mathcal{M}' &= 4T_M\mathcal{H}\mathcal{M}, \end{aligned} \tag{3}$$

have the stationary point

$$\left(\frac{1}{4T_P}, \frac{g(4T_P - 1)}{4T_P(T_P + g)}, 0\right).$$

From this the selective pressure can be found,

$$\begin{aligned} \text{selective pressure } \mathcal{S} &= \frac{\ln(4T_M\mathcal{H})}{T_M - T_P} \\ &= \frac{\ln(\frac{T_M}{T_P})}{T_M - T_P} \\ &\approx \frac{1}{T_P}. \end{aligned}$$

A new complication must now be added to our PCA model: a mutant species of parasite with a different rate of transmission must be included. For simplicity each host can be infected by at most one parasite; if two different species of parasite compete for the same host then the victor is determined probabilistically relative to the two transmission rates. The standard PCA model is run for sufficient iterations to remove transients, and then the mutant is introduced by changing the transmissibility of a small proportion of the parasites.

For the probabilistic cellular automaton, the probability of an invasion succeeding for both higher and also lower transmissibility must be taken into account. So taking $\varepsilon = |T_M - T_P|$ (the difference between the mutant and resident transmission rates) and τ as the time since introduction of the mutant, the selective pressure $\mathcal{S}$ is given by

$$\lim_{\tau,\varepsilon \to 0} \left[\frac{\mathbb{P}(\text{higher invasion})}{\tau\varepsilon} \ln\left(\frac{\mathcal{M}(\tau)}{\mathcal{M}(0)}\right) - \frac{\mathbb{P}(\text{lower invasion})}{\tau\varepsilon} \ln\left(\frac{\mathcal{M}(\tau)}{\mathcal{M}(0)}\right)\right].$$

(Note that this is consistent with the definition in the ODE case, as there the probability of higher invasion is 1.) As we are dealing with small probabilities the numerical results are computationally expensive; many simulations are necessary, but they do indicate that the selective pressure is positive up to T_c, so that there is always evolution towards the critical

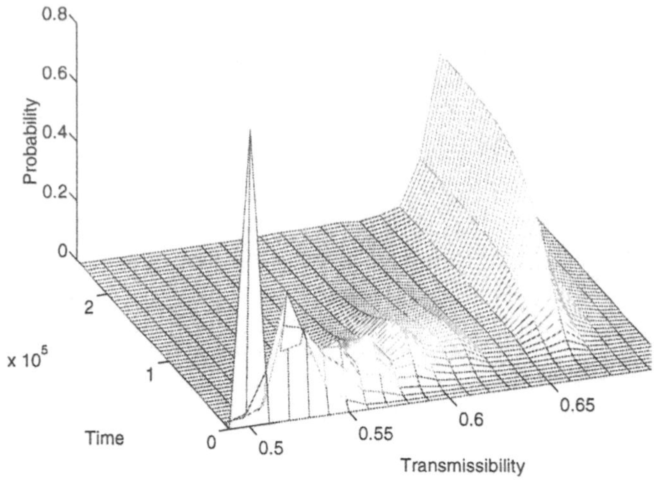

Figure 5. The convergence of a population distribution to an asymptotic limit at $T = 0.67$. This results from more transmissible parasites having the advantage locally, but those with transmissibilities greater than T_c being doomed to extinction.

point. Unfortunately complications may arise close to T_c, as the system will often collapse, so that both mutants and normal parasites die out.

Using the values of the two selective pressures ($\mathcal{S}^+(T)$ for increasing transmissibilities and $\mathcal{S}^-(T)$ for decreasing transmissibilities) a Markov process can be formed to model the behaviour of a slowly mutating population of pathogens. The probability of the resident being of transmissibility T evolves according to:

$$\mathbb{P}(T)' = (1 - \varepsilon\mathcal{S}^+ - \varepsilon\mathcal{S}^-)\mathbb{P}(T) + \varepsilon\mathcal{S}^+\mathbb{P}(T - \varepsilon) + \varepsilon\mathcal{S}^-\mathbb{P}(T + \varepsilon).$$

ϵ was chosen as 0.01, as this produced seemingly accurate results within a sensible time.

Starting with a delta function at $T = 0.51$ and iterating the map, an asymptotic distribution of probabilities is reached, with the maximal value occurring at $T = 0.67$ which is the critical transmissibility (Figure 5). One feature that is consistently found with this and other spatial models is the far lower values for selective pressure compared to the homogeneous mean-field equations. This can be attributed to the smaller number of contacts possible due to the local interactions and non-trivial spatial structure. It has long been known (Mollison, 1986) that in spatial models, as in the real world, invasions usually spread as a wave front and do not grow exponentially; this is one feature (linked to the presence of strong spatial correlations) which leads to a lower selective pressure. Another is the high risk of stochastic extinction in the initial stages of invasion due to the low numbers of mutant parasites.

3.3. DIFFERENCE EQUATIONS TO MODEL THE SYSTEM

One of our main aims is to use these PCA systems to find new methods and to obtain insights into the modelling of more realistic ecosystems. It was noticed when running the cellular automaton model that the hosts generally grew in patches, separated by empty cells (see Figure 2). These patches grew, were broken up by deaths, and amalgamated with other patches, while within the patch the parasites spread. These four basic functions were used to create the PATCH (Partitioning, Allocation, Transmission and Coalescing of Habitats) model to mimic the cellular automata behaviour by difference equations. The patches can be divided into four different habitat types,

Type 0 patches contain only hosts and empty sites.
Type 1 patches contain only hosts, empty and normal parasites.
Type 2 patches contain only hosts, empty and mutant parasites.
Type 3 patches contain hosts, empty, normal and mutant parasites.

The distribution of hosts, parasites and empty sites is assumed random within any patch. For simplicity all the patches of type i are assumed to be identical squares with area S_i^2 and they are distributed with a density d_i. The assumption that the patches are square can easily be relaxed by amending the ratio of the patch volume to patch surface area. A fractal boundary would probably be more realistic, although its dimension would have to be approximated from computation of the cellular automaton, and may be found to be a function of several variables.

The difference equation can be divided into four operations (see Figure 6), which are:

Partitioning the habitat. Given the number of empty sites in a square patch and assuming that the sites are randomly distributed an approximation to the probability that the empty sites break the patch into smaller units can be calculated. Using this we can divide up the patches producing more smaller patches.

Allocating the species. As the patches have now reduced in size it is likely that they do not contain all of the necessary species to remain in the original habitat type, for example if a patch of type 1 has been subdivided there is a good chance that not all the new smaller patches contain parasites so they will revert to type 0. By assuming a random distribution of species between the new patches they can be reallocated a habitat type.

Transmission of parasite and growth of habitat. This is the standard homogeneous mean-field approximation to the dynamics, with hosts grow-

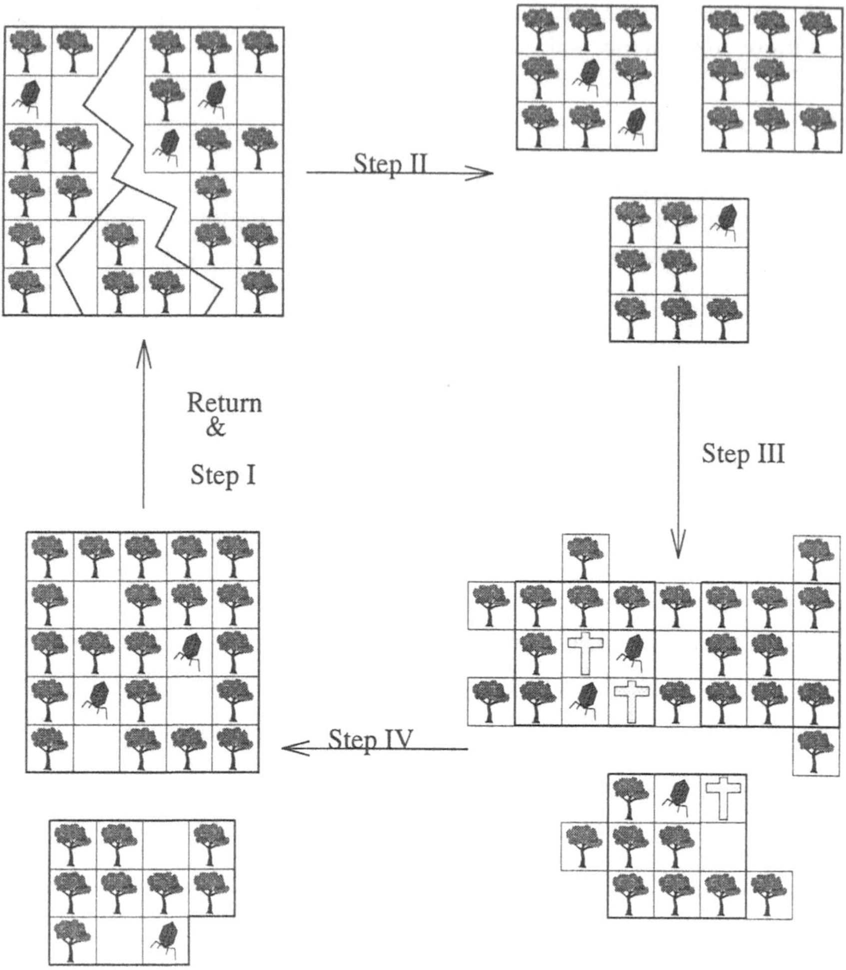

Figure 6. Pictorial Representation of the PATCH Model

ing into unoccupied sites and the parasite being transmitted to healthy hosts. The only addition is that the hosts can grow outside the patch into unoccupied territory.

Coalescing the Habitats. This growth into unoccupied areas outside the patch may have caused two nearby patches to fuse, causing a larger patch of possibly different habitat type. The entire process is now repeated.

The PATCH model does not give exactly the same results as seen in the cellular automaton, and we should not expect it to. This can be mainly attributed to the homogeneous nature within each patch, and the homogeneity of patches over space, i.e. only large scale correlations and fluctuations are taken into account. Many modifications could be made to improve the accuracy of the PATCH model, but we are searching for a simple cause for the qualitative difference between the PCA and the standard differential system.

The PATCH model has three main advantages over the cellular automaton model. Firstly it is a lot faster so that results may be obtained over a wider range of values and conditions. Secondly the PATCH model is deterministic, so that any results obtained are true for every run. This again leads to faster computation as multiple runs are no longer needed. Another advantage is its generality as it does not depend upon the precise details of the interactions and the neighbourhoods, bringing us to the conclusion that the qualitative results may hold for many host-parasite systems. Finally as the PATCH model is a set of difference equations it is possible, although highly complicated, to use standard analysis techniques. Hence it is best to use the PATCH model to get a general overview of the qualitative behaviour of the system, and only use the cellular automaton when exact numerical results are required.

3.3.1. *Results*

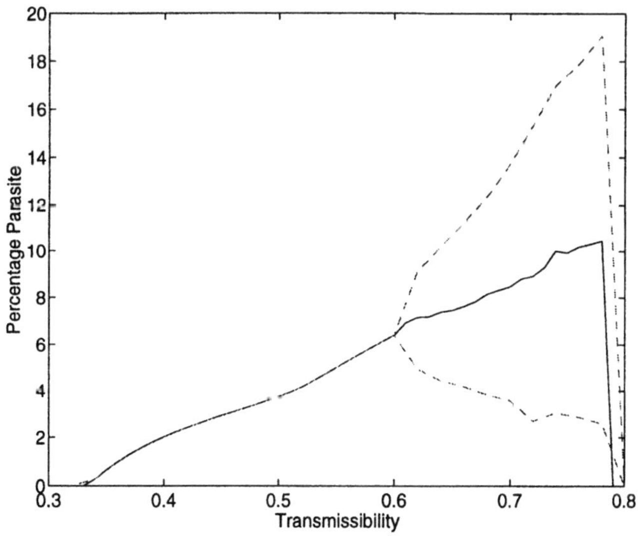

Figure 7. Percentage Parasites against Transmissibility (the PATCH Model). The solid line shows the mean, the dotted lines are the maximum and minimum values. There is a clear bifurcation at $T \approx 0.6$ and extinction of the parasite at $T > 0.8$

The cellular automata and Patch models have qualitatively similar results; both have a T_{min} and a T_c, both have an increasing percentage of parasites with transmissibility, although the PATCH model has a far sharper upper cut off.

As we increase the transmissibility the number of parasites, begins to undergo oscillations of increasing amplitude (Figure 7), which in a stochastic system may lead to extinctions at an earlier value then that predicted here. It should be noted that these oscillations would be damped in a full spatial model, it is only the homogeneity of patches which allows these dynamics. A similar behaviour however is observed in the PCA where increasing T caused the amplitude of the stochastic fluctuations about the mean to increase.

3.4. CONCLUSION

We have demonstrated that even the most simple generic caricature model for host-parasite systems has surprisingly complex and interesting dynamics when it is embedded in space. Three major features separate the results of this probabilistic cellular automaton from the usual mean field theory; the reduced range of transmissibilities over which the parasites persist; the lower proportion of parasites (by a factor of four) at all values of T and the lower selective pressure for evolution towards increased transmissibility. The results from the PATCH model lend supporting evidence to the idea that it is the spatial structure which causes these departures from the mean field theory. Due to the simplicity of this model and the highly characteristic results it will be reused several times in this chapter as a test-bed for techniques and as a basis for more complex systems.

4. Extensions to the Basic Model

The simple model introduced in the previous section can now be extended to include multiple host and parasite species, greater mobility of the hosts and more physiologically realistic evolution. This obviously requires far more parameters than before, so it will be difficult to obtain a general pattern of behaviour for the entire range of parameters values. Three areas have been chosen to highlight the difficulties and differences that arise when considering spatial models.

The first section deals with multiple hosts affected by multiple parasites, and in particular looks at the competition between two host species and the trade off between fast growth and resistance. In the second section attention is turned to the case where, due to increased mobility of individual hosts, the pathogen can be transmitted over a longer range than the usual four cell neighbourhood. In the third section the question of pathogen evolution is revisited, but applying some physiological constraints to the parameters.

Finally we return to multiple host-parasite systems for a detailed examination of the maintenance of sexual reproduction in a population.

Many of the variables used in Section 3 will need to be redefined to cope with the increased number of species. Let N_H and N_P be the number of host and parasite species respectively. A new factor, the natural death rate of the uninfected hosts $\underline{d} \in \mathbb{R}^{N_H}$ is included in the model for more realism. The vector $\underline{g} \in \mathbb{R}^{N_H}$ gives the growth rate of the uninfected hosts and this is modified by the matrix $\boldsymbol{G}(\in [0,1]^{N_H N_P})$ where G_{ij} is the factor by which the growth rate of host i is affected by parasite j. Unlike section 3 the host is now allowed to be infected by multiple parasites although this is a rare occurrence. The growth rate of host 1, infected by parasites 2,4 and 7 is

$$g_1 G_{12} G_{14} G_{17}.$$

The probabilities of transmission, virulence and recovery (back to a healthy susceptible state) associated with each host and parasite are also given by matrices $\boldsymbol{T}, \boldsymbol{V}, \boldsymbol{R}(\in [0,1]^{N_H N_P})$ respectively. When multiple infections do occur the chance of survival to the next iteration is calculated multiplicatively i.e. in the above example

$$\mathbb{P}(\text{survival}) = (1 - d_1)(1 - V_{12})(1 - V_{14})(1 - V_{17})$$

4.1. MULTIPLE HOST - MULTIPLE PARASITE SYSTEMS

The main question addressed in this section is how space influences the competition and coexistence of host and parasite species. It was seen in Section 3 that in a spatial environment the selective pressure for parasite evolution was much less than in the homogeneously mixed case. The same is true for the multi-species systems analysed here, with the time spent in co-existence increasing with the number of species in the system.

4.1.1. *Two Hosts - One Parasite*

It is first necessary to define the *equilibrium manifold* $\mathcal{M}$ for the values of growth, transmissibility, virulence and resistance associated with a particular host. Two hosts $\mathcal{H}_1$ and $\mathcal{H}_2$ will co-exist in equilibrium, if and only if

$$(g_2, T_{21}, V_{21}, R_{21}, d_2) \in \mathcal{M}(g_1, T_{11}, V_{11}, R_{11}, d_1);$$

that is, the parameters associated with host 2 belong to the manifold of host 1. Three assumptions will now be made to reduce the size of the parameter space; the death rate $\underline{d}$ is so small as to be ignored except in the absence of parasites, infected hosts are not be able to reproduce ($\boldsymbol{G} = 0$), and hosts that are not killed by the parasite will immediately recover back to a susceptible state, $\boldsymbol{R} = 1$. This still leaves three parameters $\underline{g}$, $\boldsymbol{T}$ and $\boldsymbol{V}$; we will take $g_1 = 0.05$, $T_{11} = 0.5$ and $V_{11} = 0.75$ for host 1.

Using the fact that each cell has only four neighbours it is possible to describe the local homogeneous dynamics by the mean-field equations (4).

$$\begin{aligned} \mathcal{H}_1' &= \mathcal{H}_1 + \left(\tfrac{(1-\mathcal{H}_1-\mathcal{H}_2-\mathcal{P})(1-(1-g_1\mathcal{H}_1-g_2\mathcal{H}_2)^4)}{2-(1-g_1\mathcal{H}_1)^4-(1-g_2\mathcal{H}_2)^4}\right)(1-(1-g_1\mathcal{H}_1)^4) \\ &\quad +(1-V_{11})\mathcal{P}_1 - \mathcal{H}_1(1-(1-\mathcal{P}T_{11})^4) \\ \mathcal{H}_2' &= h_2 + \left(\tfrac{(1-\mathcal{H}_1-\mathcal{H}_2-\mathcal{P})(1-(1-g_1\mathcal{H}_1-g_2\mathcal{H}_2)^4)}{2-(1-g_1\mathcal{H}_1)^4-(1-g_2\mathcal{H}_2)^4}\right)(1-(1-g_2\mathcal{H}_2)^4) \\ &\quad +(1-V_{21})\mathcal{P}_2 - \mathcal{H}_2(1-(1-\mathcal{P}T_{21})^4) \\ \mathcal{P}_1' &= \mathcal{H}_1(1-(1-\mathcal{P}T_{11})^4) \\ \mathcal{P}_2' &= \mathcal{H}_2(1-(1-\mathcal{P}T_{21})^4) \qquad \text{where } \mathcal{P} = \mathcal{P}_1 + \mathcal{P}_2 \end{aligned} \tag{4}$$

These are a more complex set of equations than equations (1) as we are now including the fact that each site experiences a limited neighbourhood, but this leads to a more robust formulation. For host 1 and host 2 to coexist it is necessary that:

$$\frac{\mathcal{H}_1 V_{11}(1-(1-\mathcal{P}T_{11})^4)}{(1-(1-g_1\mathcal{H}_1))} = \frac{\mathcal{H}_2 V_2(1-(1-\mathcal{P}T_{21})^4)}{(1-(1-g_2\mathcal{H}_2))} = M \tag{5}$$

This is the equilibrium manifold. For the host 1 parameters $M \approx 0.216$.

Returning to the spatial model and examining the question of the trade-off between a rapid growth rate ('r strategist') and resistance to the parasite ('k strategist'), it becomes necessary to define the term resistance more precisely. The case examined will be of lower virulence (the chance of dying from the disease is less); other cases such as faster recovery or lower susceptibility could be dealt with in a similar manner.

Figure 8 shows the trade-off between virulence and growth rate; graphs a and b show where the ratio of host 1 to the total number of hosts is 10%, 50% and 90% after 200 and 4000 iterations respectively. The change from dominance by host 1 to host 2 can be seen to be very smooth. The 50% contour should be compared to the equilibrium manifold (dotted line) from equation 5,

$$\mathcal{H}_2 V_{21} = M(1-(1-g_2\mathcal{H}_2)^4) \approx 4Mg_2\mathcal{H}_2 \approx 15g_2$$

This is obviously a reasonable approximation to the spatial model. It should be noted that the theoretical result is contained between the 10% and 90% contours and becomes a better approximation after successive iterations.

These results were obtained on a 150×150 lattice, where 200 iterations should be sufficient to remove any species that decays at a faster rate than $e^{-0.0466t}$ and 4000 iterations should remove any species that decays faster than $e^{-0.00233t}$; yet co-existence is seen over a fairly wide parameter range.

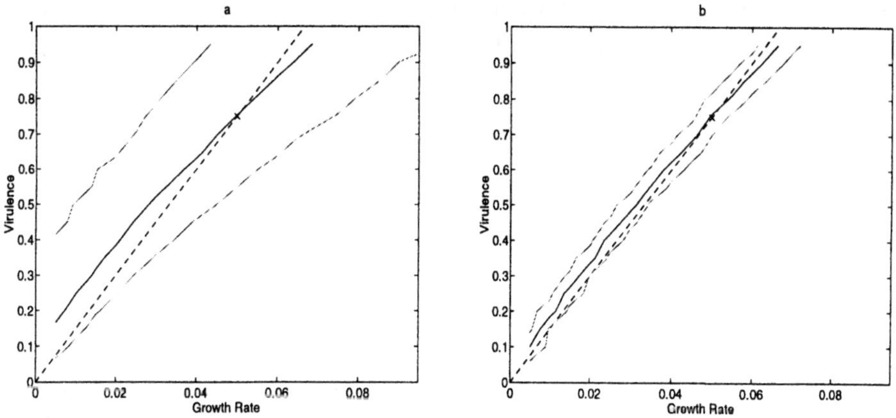

Figure 8. The competition between to host species, showing the trade off between virulence and growth rate. Graph *a* is after 200 iterations, graph *b* is after 4000 iterations. The solid lines on the graphs are the 10%, 50% and 90% contours , the dotted line is the theoretical equilibrium manifold predicted by the mean field equations.

Again we see far slower selective pressure in the spatial model as competition between hosts can only occur along a narrow boundary; most of the hosts are only in competition with other hosts of the same kind. The main question that arises now is why the homogeneous model should be a reasonable approximation for host behaviour but not for parasite behaviour.

Each parasite has a very short life time (usually one iteration) and so experiences relatively large scale fluctuations and this is seen in the stochastic nature of the results. Its transmission is very dependent on its local environment and hence the spatial correlations formed in the PCA. These features are not modelled by the homogeneous equations. The hosts, however, live for far longer and so experience a neighbourhood which is far more of a temporal average, hence the results obtained are smoother. When competition between hosts occurs it is always in a narrow band, usually only a few sites across, and so both host types experience a similar proportion of parasites despite the large scale spatial structures. Hence the results more closely match the mean-field results.

4.1.2. *Two Hosts - Two Parasites*

If another layer of complexity is added to the system, by having two parasites and two hosts then a richer set of behaviours is found. Co-existence of both host species and both parasite species is now possible (as predicted by the mean field equations), and this gives rise to far more complex dynamics. Define $\mathcal{H}_i$ to be the proportion of cells occupied by a healthy host i, and let $\mathcal{P}_{i,j}$ be the proportion of cells with a host i infected with parasite j. The proportion of hosts infected with both parasites is very low and therefore will be neglected in formulating the mean-field equations.

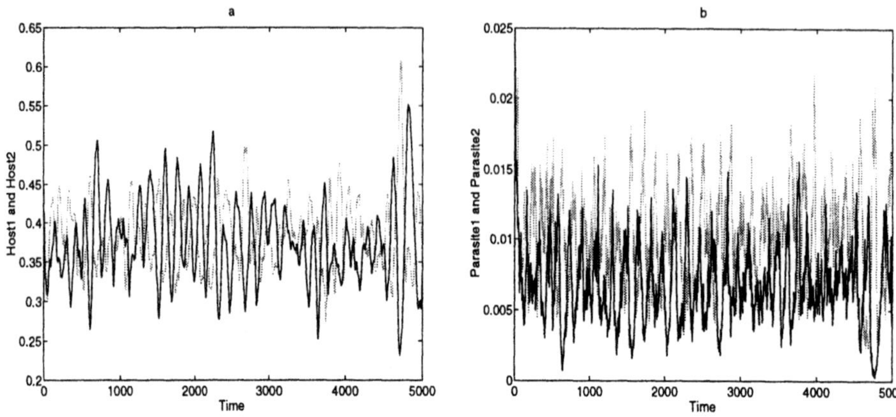

Figure 9. Graph showing chaotic oscillations for the two host species (a) and the two parasites species (b). It should be noted that the dynamics of the hosts are far less stochastic than those of the parasite due to the longer life spans and slower growth rates.

As an example of the behaviour seen, the following system will be examined,

$$\underline{g} = (0.05, 0.05) \qquad \underline{d} = (0,0) \qquad \boldsymbol{G} = \begin{pmatrix} 0 & 0 \\ 0 & 0 \end{pmatrix} \qquad \boldsymbol{R} = \begin{pmatrix} 1 & 1 \\ 1 & 1 \end{pmatrix}$$

$$\boldsymbol{T} = \begin{pmatrix} 0.8 & 0.2 \\ 0.2 & 0.8 \end{pmatrix} \qquad \boldsymbol{V} = \begin{pmatrix} 0.7 & 0.8 \\ 0.7 & 0.8 \end{pmatrix}.$$

The results were obtained on a 100×100 lattice, which (using the method for non-stationary systems developed by Keeling *et al.*, 1997a) is the coherence length scale. The spatial system is chaotic (this was checked by calculating the Lyapunov exponent) and possesses large amplitude oscillations with periods of around 130 iterations (figure 9) the mean field equations which are merely the two-parasite extension of equations (4) produce simple cycles with a period of around 40 iterations. This difference in behaviour types is not limited to these particular parameters. A Monte Carlo search was made through the four dimensional space of variables keeping $\boldsymbol{T}$ symmetric and the rows of $\boldsymbol{V}$ equal. Out of 10,000 variable choices all of the mean field simulations exhibited either cyclic behaviour (although often quite complex) or damped oscillations tending to a fixed point, so the chaotic behaviour appears to be due solely to the individual nature and spatial framework.

4.1.3. *Multiple Hosts - Multiple Parasites*

We shall now consider the case of a system with a very large number of host and parasite species; obviously to arbitrarily asign values to the matrices $\boldsymbol{G}$, $\boldsymbol{R}$, $\boldsymbol{T}$ and $\boldsymbol{V}$ would be meaningless. Instead we will assume that each host has a genotype G^H (lying between 0 and 255) which is a string of eight binary bits $(G^H_1, \ldots, G^H_8)$. Each parasite also has a genotype G^P which again is an eight bit string, the closer the bits of this string matches the host string the higher the parasite's transmissibility T and virulence V and the lower the host's recovery R and growth rate g. To continue the theme of growth rate versus resistance and to see if parasitism could maintain genetic diversity, the growth rate increases with the genotype of the hosts (controlled by g_G) so that in the absence of parasitism some selective pressure still exists. The growth, transmissibility, virulence and recovery rate are given by:

$$\begin{aligned} g &= \left(g_L + g_M M(G^H, G^P)^2\right)\left(1 + g_G G^H/255\right) \\ T &= \left(T_L + T_M M(G^H, G^P)^2\right) \\ V &= \left(V_L + V_M M(G^H, G^P)^2\right) \\ R &= \left(R_L + R_M M(G^H, G^P)^2\right), \end{aligned}$$

where $M(G^H, G^P)$ is the proportion of sites in the two strings that are the same; this value is squared so that there is a far greater weighting for a perfect match. This is equivalent to having the parasites specialising on particular host types. If the host is parasitised by two or more parasites then M is replaced by:

$$1 - \left\{1 - M(G^H, G^{P_1})\right\}\left\{1 - M(G^H, G^{P_2})\right\}\ldots\left\{1 - M(G^H, G^{P_n})\right\}$$

so that the effects of parasitism are enhanced. In fact multiple parasitism is rarely encountered.

Unfortunately, a vast lattice would be needed if we wished to maintain all of the 256 host and 256 parasites types and allow them to compete and interact fully. The only means of ensuring complete survival on a feasible sized lattice is to include regeneration of extinct genotypes, the two obvious and biologically realistic methods of accomplishing this are sexual recombination and mutation. We will allow the hosts to 'breed' with a randomly selected host from its local environment and the genotype of the off-spring is a random combination of the parents' genotypes. To allow the parasites to respond to a changing host environment their genetic string can slowly mutate, the probability of a change at any one site is μ per iteration. All of the variables used are given in the table below.

d	g_L	T_L	V_L	R_L	g_G	g_M	T_M	V_M	R_M	μ
0.005	0.02	0.1	0.0	1.0	1.0	-0.014	0.7	0.6	-0.9	0.1

Initially the hosts with high genotype numbers have a considerable advantage, up to two fold, over the other hosts and their numbers increase. This increase in high genotypes gives a reproductive advantage to parasites with high genotypes, which in turn reduces the advantage possessed by the high genotype hosts. This negative feedback is made all the more effective as there is some positive correlation between local genotypes. Figure 10 shows the long term average percentage for each bit being set: there is little variation in the values once transients have been eliminated. It can be clearly seen that there is a slight advantage in possessing any given bit set, but this effect is by no means as strong as observed for the homogeneous mean field equations. It should also be noted that there is far more advantage in having the higher bits set, this is presumably because they confer a greater increase in growth rate for a comparable increase in parasitism. It appears that, together with sexual reproduction, parasitism can have a strong enough effect to maintain diversity within the population.

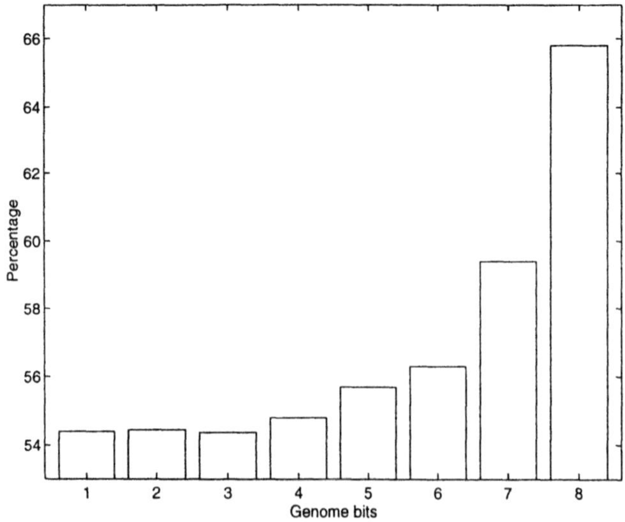

Figure 10. Graph showing the average percentage of genotypes in which each bit is set. Two facts should be noted: firstly that there appears to be a slight advantage to having any bit set and secondly that the advantage increases as the effect of the bit increases.

We shall return to this model at the end of the section to see how it can explain the persistence of sexual reproduction in a population, due to parasitism and the diversity of a sexual species.

All of the models so far described suffer from the same problem: each site is only affected by its closest four neighbours, so correlations only form initially on the most local of scales. There will be three models addressed here, each one based on the simple host-parasite PCA, but representing a different way of interpreting the change from a fixed four cell local neighbourhood to a larger one. In the first model each site is coupled to all the other sites in the increased neighbourhood but the transmissibility and growth rates are reduced so that the expected growth or transmission from a site remains constant. In the second and third examples each cell is only coupled to four sites but these are chosen randomly from the N cells of the neighbourhood. In the second case the sites are chosen randomly at each iteration; in the third example the sites are chosen at the start of each simulation and then remain fixed.

It is often conceptually easier to consider the growth and transmission to be occurring over a longer range r, a diamond shaped neighbourhood will be used to allow smooth transition from the four cell Von Neumann neighbourhood ($r = 1$). The number of neighbours associated with a range r is:

$$N_r = 2r(r+1)$$

When coupling to all of the N cells in a neighbourhood the transmissibility T_N and the growth rate g_N are taken as:

$$T_N = \frac{T^*}{N} \qquad \text{and} \qquad g_N = \frac{g^*}{N}$$

As N becomes large the system becomes globally coupled and the following homogeneous mean-field equations hold:

$$\begin{aligned} \mathcal{H}' &= \mathcal{H} + (1 - \mathcal{H} - \mathcal{P})\left(1 - (1 - g_N\mathcal{H})^N\right) - h\left(1 - (1 - T_N\mathcal{P})^N\right) \\ \mathcal{P}' &= (1 - V)\mathcal{P} + \mathcal{H}\left(1 - (1 - T_N\mathcal{P})^N\right). \end{aligned}$$

In the limit $N \to \infty$ and with $V = 1$ this is comparable with the Nicholson-Bailey equations (Nicholson & Bailey, 1935) with density dependent growth of the hosts:

$$\begin{aligned} \mathcal{H}' &= \mathcal{H}\left(e^{-\mathcal{P}T^*}\right) + (1 - \mathcal{H} - \mathcal{P})\left(1 - e^{-\mathcal{H}g^*}\right) \\ \mathcal{P}' &= \mathcal{H}\left(1 - e^{-\mathcal{P}T^*}\right). \end{aligned}$$

The change in behaviour associated with an increasing range is illustrated in Figure 11. At short ranges the results show the large amount of stochastic variation and the lower parasite burden which characterised the model in Section 3. As the range increases the average number of parasites quickly approaches the value predicted by the homogeneous equations, attaining 90% of its value at a range of only three cells. For larger ranges still the stochastic variations decrease and are swamped by decaying oscillations of a similar amplitude and period to the homogeneous equations. There is still insufficient global coupling however to remove all fluctuations and the inherent stochastic nature of the system introduces further oscillations. If the range is set equal to half the lattice size (75 cells) then the homogeneous equations are a very good approximation to the model.

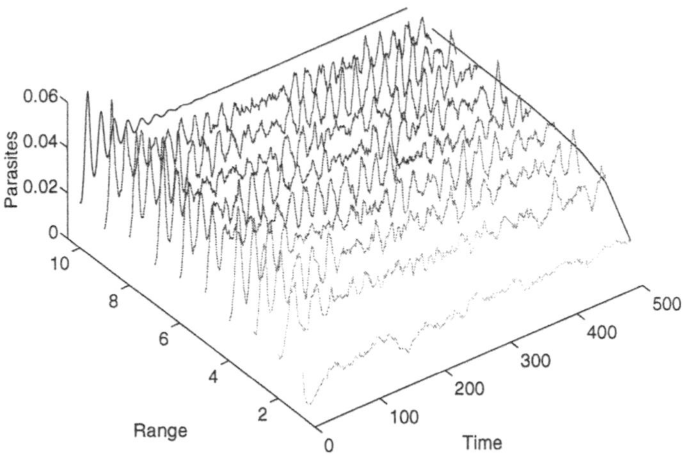

Figure 11. Change in the parasite behaviour associated with greater ranges of infection. Grey lines are for the simulations, the black line is the homogeneous mean field solution, the line at *time* = 500 gives the average number of parasites over the simulation. $T^* = 2.0$ and $g^* = 0.2$ ($T_1 = 0.5$ and $g_1 = 0.05$)

Why should such a change in behaviour be observed? Two main reasons for this are (i) the greater stochasticity associated with small neighbourhoods, and (ii) the change in coupling times between neighbouring sites.

The changes in stochasticity are fairly straightforward. For small neighbourhood sizes the random change of one site has a relatively large effect, whereas, when the neighbourhood is larger, this effect is reduced. If the sites in the neighbourhood are assumed normally distributed, then as the neighbourhood increases so the ratio of the standard deviation to the mean decreases. The change in coupling is slightly more complex: given a centre site and one other in its neighbourhood then as the neighbourhood increases the central site exerts less and less effect on its neighbour (as both T and g scale like $\frac{1}{N}$) and so the coupling between sites decreases.

When picking four neighbours at random from within a set of N neighbours the connections between sites were taken as unidirectional. This means that the behaviour of each cell can be influenced by its four neighbours but the converse is not true. The space is therefore a random digraph of degree four and hence is almost always weakly connected (Bollobás, 1985). This means that an infection can usually spread to all parts of the system. If each site were connected to only four others and the connections were in both directions, then not only would the calculation of such a network require vast amounts of computation but the resulting graph would not usually be connected. Due to the one directional nature of the interactions there is no significant difference between choosing the sites randomly at each step or choosing the sites initially except in terms of computational speed.

In general it is only the stochasticity that differentiates between any of the methods of extending the range of infection and growth. In situations where both the parasite and host are sufficiently long-lived so that they can observe a temporal average, there is very little quantitative difference between the three techniques.

This method of extending the neighbourhood gives us a means of interpolating between homogeneous mean field dynamics and PCA. It highlights the importance of stochasticity and spatial correlations to the dynamics of a system and clearly shows why we can expect creatures that interact with a relatively large habitat to exhibit dynamics close to the mean field approximations.

4.3. EVOLUTION WITH PHYSIOLOGICAL CONSTRAINTS

In Section 3, it was assumed that the pathogens can mutate freely and independently over the two dimensional space of virulence and transmissibility. However this is clearly not the case, as the transmissibility and virulence of a pathogen are determined by physiological mechanisms. For a better understanding of the selective pressures on the host-pathogen system the underlying physiology must be modelled.

4.3.1. *Physiological Modelling*

We shall form a very simple model for the interaction between the immune system and an invading pathogen, and let this model influence the mutation of transmissibility and virulence in the PCA. Let P be the density of pathogen particles and I be the density of antibodies in response from the immune system.

$$\begin{aligned} \dot{P} &= \alpha P - \phi I P \\ \dot{I} &= \gamma P \end{aligned} \tag{6}$$

this can be written as,

$$\ddot{P} = \frac{\dot{P}^2}{P} - \phi\gamma P^2 \tag{7}$$

which reduces the number of variables that have to be considered as $\phi\gamma = \beta$ can be treated as a single constant. The results from equation (7) are given in Figure 12 and a fairly realistic curve is obtained considering the naive simplicity of the model. The transmissibility T and virulence V of the pathogen in a single iteration can be determined in terms of the number of particles and will be taken as the infinite limit of the following:

$$\begin{aligned} T(t) &= 1-(1-T_0)^A \quad \text{where} \quad A = \int_0^t P(s)(1-V(s))ds \\ V(t) &= 1-(1-V_0)^B \quad \text{where} \quad B = \int_0^t \Theta(P(s)-\delta)ds \end{aligned} \tag{8}$$

and Θ is the Heaviside function. Obviously this is a vastly over simplified caricature model of the immune system response to an invading pathogen; it is hoped however that it will capture many of the salient features.

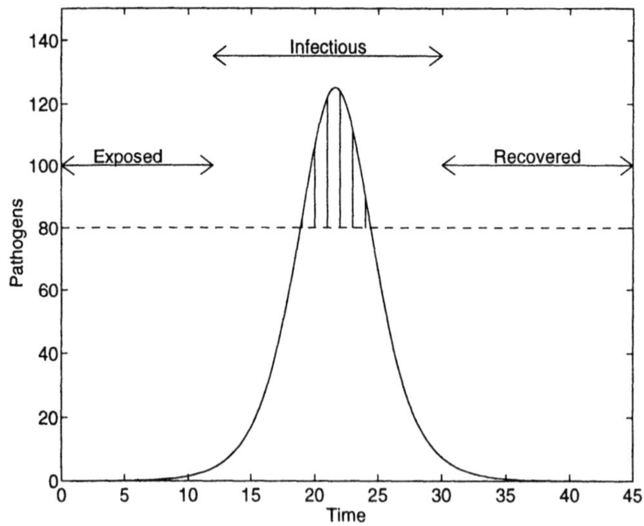

Figure 12. The growth and decay of the density of pathogens within a host. The total area under the curve gives a measure of the transmissibility, the shaded area measures the virulence.

It will be assumed that mutation in the growth rate α and the effect of the immune system β will be easier and more prominent than changes in the other parameters. The other parameters δ, T_0 and V_0 can be thought of as either being defined by the host or as being dependent on the pathogen's methods of transmission and reproduction. Altering α or β while keeping the other parameters constant will produce trajectories in the transmissibility-virulence plane which successive mutations should follow; the shape of these trajectories determines the long term behaviour of the system.

If the pathogen is assumed non-virulent (i.e. if δ is large) then the value of A as $t \to \infty$ can be calculated.

$$\begin{aligned} \ddot{P} &= \frac{\dot{P}^2}{P} - \beta P^2 \\ \Rightarrow \qquad P &= -\frac{1}{\beta}\left(\frac{P\ddot{P} - \dot{P}^2}{P^2}\right) = -\frac{1}{\beta}\frac{d}{dt}\left(\frac{\dot{P}}{P}\right) \end{aligned}$$

Hence the integral for A is,

$$A = \int_0^t P ds = -\frac{1}{\beta}\left[\frac{\dot{P}}{P}\right]_0^t = \frac{1}{\beta}\left(\alpha - \frac{\dot{P}(t)}{P(t)}\right)$$

The limit as $t \to \infty$ in the above equation cannot easily be calculated; however it can be seen that when at the point of maximum infection ($t = t_{max}$), equation (7) is identical in forward and reverse time.

$$A = \frac{2\alpha}{\beta} + \int_{2t_{max}}^{\infty} P(s)ds$$

If it is assumed that $P(2t_{max}) \equiv P(0)$ is small then the final values of P can be approximated by a decreasing exponential (by setting $P^2 = 0$ in equation (7) and using the relation $I = \gamma A$). So,

$$A \approx \frac{2\alpha}{\beta} + \frac{P(0)}{\alpha}$$

This means that for all reasonable choices of parameters and starting conditions the transmissibility T is an increasing function of the growth rate of the pathogen and a decreasing function of the immune response, providing the effects of virulence are ignored. The virulence V is always an increasing function of α and a decreasing function of β, as B (in equation (8)) is a simple integral.

When the virulence is incorporated into the equations the changes in transmissibility become more complex as there is now a chance that the pathogen has killed the host before transmission has occurred. We will now use equation (8) to shape the mutation of the transmissibility and virulence by giving the pathogen a fixed mutation rate in the parameters α and β, i.e. we will couple the notion of evolution to critical transmissibility with a more realistic parameter space. Two cases will be examined which demonstrate two distinct behaviours due the the effects of virulence.

4.3.2. *Evolution of the System*
In both simulations $T_0 = 0.02$, but the effects of virulence are different with the variables being set as follows: for simulation 1, $\delta = 600$ and $V_0 = 0.2$; for simulation 2, $\delta = 100$ and $V_0 = 0.04$.

The variables α and β were then free to mutate, and the results obtained are shown in figure 13.

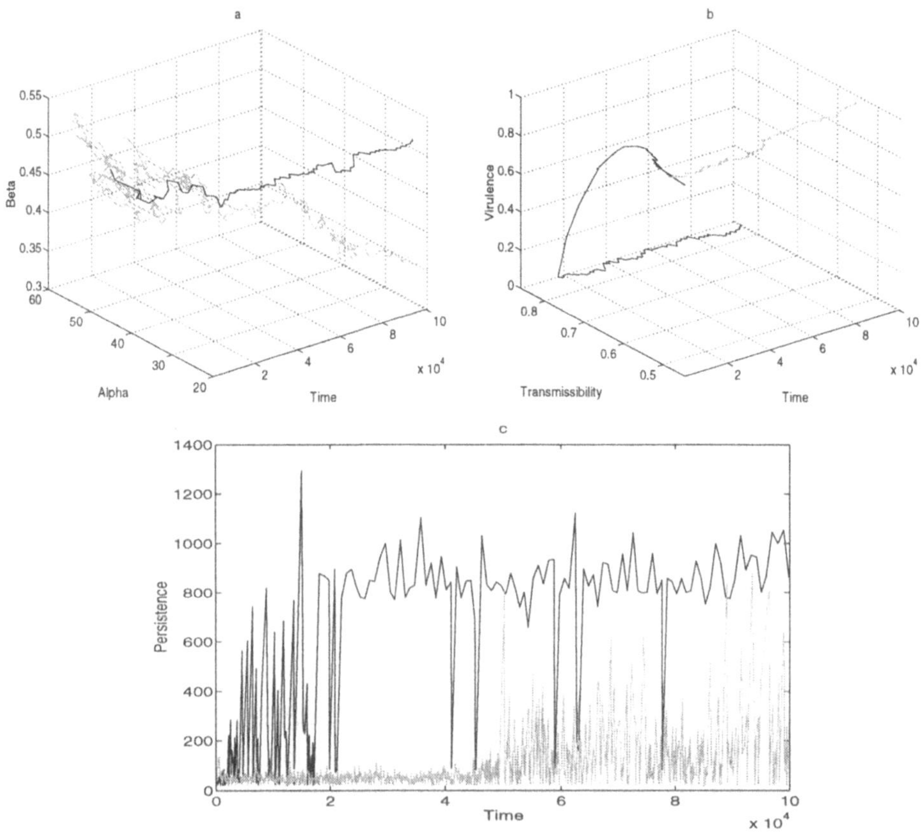

Figure 13. Graph *a* shows the evolution of α and β, graph *b* shows the associated values for transmissibility and virulence and graph *c* shows the persistence of the phenotype. Black is for simulation 1, grey for simulation 2.

After the transient dynamics died out, a mutant parasite was added as in section 3. The α and β parameters of the mutant lie within 0.01

of the parameters of the resident parasite. A subsequent mutant is only added when either the previous mutant or a resident has become extinct, so there are at most two species of parasite at any one time. In graph *a* it can be seen that for simulation 1 the parasite evolved to a higher value of β but a similar value of α compared to simulation 2. This translates into the transmissibility and virulence in graph *b*, which is equivalent moving between the two different 'phenotype' spaces. In simulation 1 where V_0 is large (so that there is a rapid increase in virulence above the threshold δ) the virus quickly evolves to zero virulence and a large transmissibility of around 0.9. In simulation 2 where the increase in virulence starts at a lower threshold, but does not increase as quickly, the virus evolves to have a virulence closer to one, and a transmissibility of around 0.55, which compares to the critical transmissibility of Section 3. It should be noted that the first simulation gives rise to a far more stable configuration, with around 800 iterations between successful invasions. Figure 14 shows an example of the expected dynamics of the disease within a host for the two simulations.

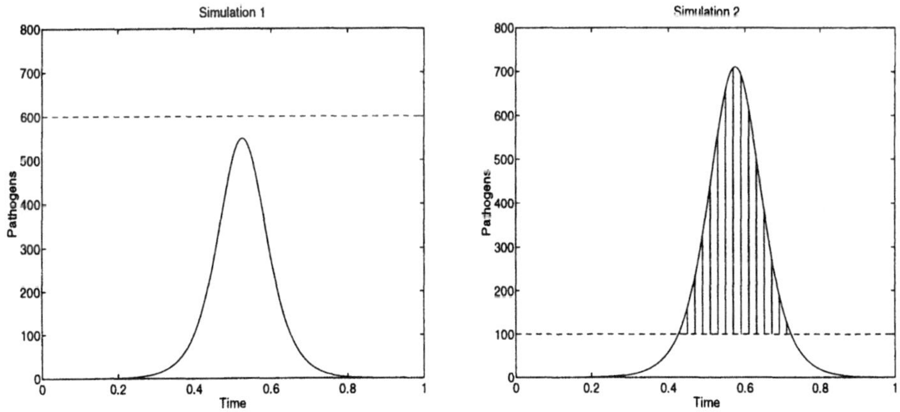

Figure 14. The dynamics of the disease for the final states of the two simulations. Again the area below the curve is related to the maximum transmissibility when the host survives infection; the shaded area above the δ-line gives a measure of the virulence.

In simulation 1, the penalty for crossing the virulence threshold δ is very severe, and the reward in extra transmission small in comparison; hence it evolves to a state where there is maximal transmission but no virulence. In simulation 2, the penalty for crossing the threshold is much less, and the reward large in comparison to the transmissibility below the threshold.

4.4. THE PERSISTENCE OF SEXUAL REPRODUCTION

The persistence of sexual reproduction in a population has long been a problem in theoretical ecology. In 1862 Charles Darwin observed that "*We do not even in the least know the final cause of sexuality; why new beings should be produced by the union of the two sexual elements, instead of by a process of parthenogenesis ... The whole subject is as yet hidden in darkness.*" Over the next century the theory of the subject progressed slowly and in 1976 J. Maynard Smith stated that "*One is left with the feeling that some essential feature of the situation is being overlooked.*" The problem may be simply stated; a parthenogenic (thelygenic) female may produce twice as many daughters as a sexual (amphimict) female because males are redundant, and yet sexual creatures abound. What is the factor that can overcome this 2:1 advantage? The fact that sexual reproduction can lead to a speeding up of evolution and the clearance of deleterious mutations may play a partial role. The most promising answer however, was that sexual recombination maintains genetic diversity, which is vital for avoiding large scale epidemics from highly specialised diseases and parasites (Levin, 1975; Maynard Smith, 1976; Bremermann, 1980). Evidence for this comes from the fact that sexual species tend to occupy undisturbed, complex habitats where parasitism is common (Glesener & Tilan, 1978), these environments are the hardest to simulate using standard modelling techniques which may explain why many of the early systems failed to predict the persistence of sex.

Solid ecological examples of increased parasitism causing an increase in the ratio of sexual to asexual reproduction are rare. It requires a species which can reproduce both sexually and asexually and inhabits many distinct environments identical except for the number of parasites. Two such examples exist, one is *Bulinus truncatus*, a fresh water snail from Nigeria where it has been shown there is a positive correlation between parasite prevalence and sexual reproduction, with higher numbers of the sexually reproducing species in lakes with a large number of parasites (Schrag *et al.*, 1994). The other example is the Topminnows (genus: *Poeciliopsis*) which are found in desert pools of the American Southwest. It has been observed that the most common clone in any pool is the most infected by flatworms, and that the sexual species has by far the lowest average burden (Vrijenhoek, 1993). In pools where the sexual population is derived from just a few individuals and so the genetic diversity is low, the sexual species is equally infected by the flatworm, showing that it is the diversity that is important.

Secondary evidence can be gained by looking at the effects of diseases on the large monocultures of cultivated crops. During the 1970's the grassy-stunt virus devastated rice crops from India to Indonesia due to genetic similarity of the crops, but now the cultivated type has been bred with *Oryza nivara* and a resistant hybrid is now grown across 110000 km^2. A similar exemplary story comes from the coffee plantations in Brazil. Most of the coffee plants are descended from a single tree from Kaffa in south west Ethiopia. In 1970 coffee rust spread though Brazil with alarming rapidity threatening many farmers with bankruptcy. Again, wild resistant strains came to the rescue (Wilson, 1992).

One possible explanation for the persistence of sexual reproduction is based on the "Red Queen Hypothesis" (see Bell, 1982). The basic idea behind this is very simple: any parthenogenic clone will have a near two fold advantage when rare. However, as the clone becomes more common there is increasing selection for parasites to be able to infect it.

Indeed, however fast hosts evolve to avoid parasitism, parasites will evolve just as quickly to be able to overcome the defences (the evolution of drug resistant bacteria or of pesticide resistant pests are good examples). This leads to an "arms race" between hosts and parasites with the parasites constantly chasing the hosts but remaining a distance away; in *Alice in Wonderland* the Red Queen tells Alice that "You have to keep running just to stay where you are." This is also true for the hosts and parasites, which must keep evolving just to survive.

This idea for the maintenance of sexual reproduction in a population has been the subject of much work recently (Hamilton, 1980; Hamilton *et al.*, 1990; Lively & Howard 1994). All these approaches used individual-based models, and assumed homogeneous mixing in the population; this has the disadvantage that a genotype has to be highly abundant in the population before it is subjected to specialisation and high levels of parasitism. In the spatial model described below, a genotype only has to be locally abundant to suffer parasite selection, so a smaller proportion of the total population is necessary for this to take effect. Spatial systems also have lower selective pressure and slower invasions than their mean field counterparts (as shown in Section 3): this will benefit sexuality.

The two models that are given below rely on a more general mechanism by which sex can resist parasites by creating genetic diversity in the host population. This removes the need for the dynamic cycling which is is necessary for Hamilton's mechanism. Spatial structure can replace the temporal heterogeneity and localised randomness can be substituted for deterministic cycling of genotypes. The models are based upon the four following premisses:

i) Both the sexuals and the asexuals live in strongly spatially correlated habitats, that is both species are assumed to exist in patches. Asexual patches are genetically homogeneous as they are formed by identical clones. Sexual patches however display great genetic diversity even at the local scale due

to the new genotypes produced by recombination.
ii) Rapidly reproducing parasites may slowly evolve by random mutations to specialise on the local host environment. In a uniform environment, as provided by the asexual patches, parasites can specialise more readily leading to far greater transmission rates and lowering the host's reproductive fitness.
iii) Due to the spatial-embedding and the short range interactions of both hosts and parasites it is only along the thin boundaries between patches that the hosts can compete and the parasites cross from one population to another. This has two main consequences; the numbers of an invading species are limited by the speed with which it can spread, and a small uninfected asexual population will maintain its two fold advantage until it becomes infected which requires a parasite to cross the boundary.
iv) The system is assumed to be at a stationary equilibrium, with local fluctuations caused by the stochastic nature of the model. The cycles that are such an important feature of Hamilton's model are not present at any scale in our model.

4.4.1. *A Probabilistic Cellular Automaton Model*

To test the hypothesis that it is parasites that maintain the need for sexual recombination, a cellular automaton simulation was built (as pictured in Figure 15).

The multiple host - multiple parasite model was refined to include both sexual and asexual individuals. The sexual species will proceed as described before (§4.1.3), the asexual species will differ in the fact that its growth rate g is increased by a factor of two and its offspring are identical clones. Later, to allow a greater amount of plasticity in the model, the sexuality rate S will be introduced, this is the proportion of the time the host breeds sexually. An individual with sexuality S will have a growth rate $(2 - S)$ times faster than a purely sexual individual; in this way we can examine gradual changes from sexuality to asexuality. The parasites could also be allowed to adapt by sexual reproduction; although this would decrease the number of parasite generations before a host could be fully exploited, it would increase the complexity of the model and add another shorter time scale. In balance it was decided to model the adaptation by mutation (reminiscent of short-lived viral diseases) rather than by sexual reproduction (as seen in longer lived parasites).

4.4.2. *Competition between Sexuals and Asexuals*

Primarily, interest will be centred on the invasion of sexual hosts by asexual ones (i.e. competition between $S = 0$ and $S = 1$). As can be seen in figure 15 the asexual hosts form clumps of identical genotypes, which are then

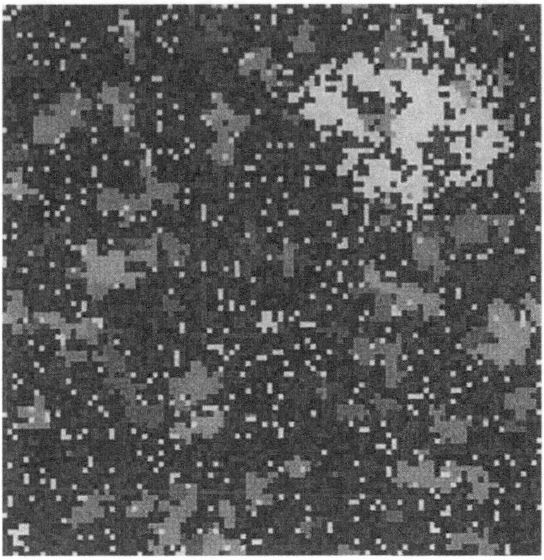

Figure 15. A typical pattern of sexual (dark grey) and asexual (light grey) hosts. Parasites (black) have just destroyed a large patch of asexuals in the top right.

vulnerable to a specialised parasite, whereas the sexual hosts are far more diverse and hence some are less susceptible. This idea agrees with the earlier assertions and will now be quantified.

Figure 16 shows the proportion of sexual and asexual hosts at various mutation rates μ for the parameters given in section 4.1.3.

Similar results are obtained for weaker effects of parasitism (i.e. g_M, T_M, V_M and R_M are of less magnitude) but larger grids, more iterations and many more simulations would be needed to get results of similar clarity. The starting conditions are for a low density of both host types, so that the initial two-fold advantage of the asexuals does not swamp the system.

Similar results are obtained independent of the initial conditions so long as the sexual hosts survive long enough for the parasites to become well established. For large values of μ the parasites are able to mutate so fast that they quickly evolve in response to their local environment and so can take advantage of both sexual and asexual hosts. This is biologically unstable, if the parasites evolved fast enough that they could always take advantage of any host then the host species would no longer be viable and there would be selective pressure towards hosts with shorter life cycles.

For lower values of μ the parasites are only able to specialise if there are areas of identical hosts as only then is there enough time for the sufficient number of mutations. This means that only patches of identical asexual hosts are badly affected by the parasites and these are quickly wiped out, again we find the presence of strong local correlations and the random en-

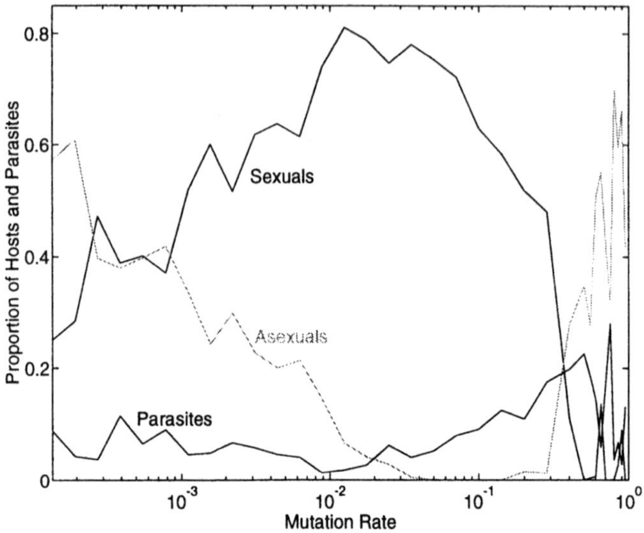

Figure 16. A graph of the number of sexual and asexual hosts against the mutation rate of the parasite μ. The numbers are from repeated simulations and are a 500 iteration average after 500 iterations have been ignored due to transience.

vironment experienced by a parasite leading to the differences between the PCA and the standard mean-field results. As the value of μ decreases still further, larger and larger numbers of identical hosts are required before the parasites become specialised. This has two effects, firstly small patches of asexual hosts will escape highly detrimental parasitism so there will always be a number of asexuals in the population, and secondly as μ decreases even further the area needed becomes comparable with the grid size and the sexual population is forced out. When the system was run on a larger grid for longer the sexuals dominated for a larger range of μ.

4.4.3. *Typical Population Dynamics*

Figure 17 is from a typical simulation with $\mu = 0.1$. It clearly shows a rapid increase in the number of asexual hosts at the early stages as they have a two fold reproductive advantage over the sexual hosts. However the high concentration of genetically identical hosts after 50 iterations causes an increase in parasite numbers which rapidly reduce the sexual population. Now that the areas of asexual dominance have a high parasite burden, the less vulnerable sexual population is able to out-compete it and the sexual numbers slowly rise. After the initial peak in the number of parasites at around 130 iterations the population settles down to a fairly constant value, despite the decrease in the asexual population.

These results do not agree with the usual non-spatial models, which

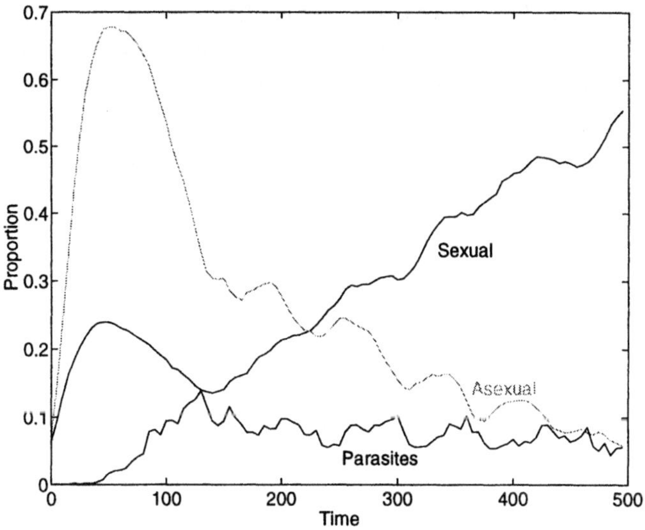

Figure 17. A graph showing the change in host and parasite numbers over time. A rapid growth of asexuals is seen before the parasites are able to specialise sufficiently.

predict that the genotypes of hosts and parasites should cycle each other; this simulation predicts a statistically constant distribution of genotypes and relies on spatial rather than temporal heterogeneities. Hence, although towards asexual hosts the system behaves according to the Red Queen Hypothesis with parasites specialising on the most common genotype, for the sexual population the model shares many similarities with the Tangled Bank model as described by Bell (1982), where the advantage of a diverse population is in coping with a heterogeneous environment. Bell puts forward many ecological ideas that demonstrate that sexual populations most closely follow the Tangled Bank model.

The results presented above are all for the competition between a population of purely sexual individuals and a population of purely asexuals. However, it is more natural to allow the sexuality rate S to slowly evolve because genetic variability has been found in populations for the frequency of recombination. Also it is believed that the re-emergence of asexuality in many of the higher organisms evolved from sexual reproduction due to females producing a small proportion of unreduced automitic eggs. The sexuality rate of the offspring of two parents with sexuality rates S_1 and S_2 is $\frac{1}{2}(S_1 + S_2)$. Three new scenarios of invasion and evolution have been studied.

For the range of mutation rates μ considered, if two species which are nearly equal in sexuality (with sexuality rates S and $S - \delta S$) compete from equal starting conditions, then the lower sexuality rate almost inevitably wins, and hence a slow decrease to asexuality is seen. Equal starting conditions is highly unrealistic but this does illustrate that we would not expect to see the emergence of sexuality from asexuality by slow evolution of S.

If however the system starts with the entire population being of sexuality S close to 1 and only a small number of mutants $S-\delta S$ are added (which is a more realistic scenario), then for $\mu = 0.1$, the population eventually returns to full sexuality as the mutants fail to invade.

Finally, if the initial state is for all individuals to start with a low sexuality rate ($S < 0.5$) then by mutation a slow decline to asexuality is observed, with the selective pressure decreasing as the population approaches pure asexuality. Thus these results suggest that in the real world we should find that both sexual and asexual populations are stable to small populations with different sexualities; although changes in the parasitism of a species could drastically alter the selective pressures on S and changes in sexuality can result.

4.4.4. *A Simple Differential Model*

Based on the observed behaviour and results from the cellular automaton (§4.4.1) a new, far simpler set of differential equations will be constructed and analysed. Let $\mathcal{H}$ and $\mathcal{P}$ refer to healthy and parasited hosts respectively, with subscripts s and a to differentiate between the sexual and asexual populations. The precise genotypes will be ignored, and instead it will be assumed that due to local genetic variation the transmission of infections to and from sexual individuals is lowered by a factor ρ. This lower transmission is indicated by the lower proportion of infections in the sexual population as seen in the PCA model. The spatial patches which were such an important factor in the previous models will be simulated by a parameter ε which measures the aggregation of the two populations. $\varepsilon = 1$ corresponds to total mixing of the populations and $\varepsilon = 0$ is when there is no parasite transmission between them. For this work ε will be taken to be small. Assuming the ratio of susceptible to infected within an neighbourhood is the same as the global density, the proportion of infected individuals a sexual host experiences is,

$$\frac{\mathcal{P}_s + \varepsilon \mathcal{P}_a}{(\mathcal{H}_s + \mathcal{P}_s) + \varepsilon(\mathcal{H}_a + \mathcal{P}_a)}$$

With logistic growth of the hosts (carrying capacity k), no recovery from infection, and constant virulence due to any infection, the following equations are obtained:

$$\begin{aligned}
\frac{d\mathcal{H}_s}{dt} &= g\mathcal{H}_s\left(1-\frac{(\mathcal{H}_s+\mathcal{H}_a)}{k}\right)-T\rho\left(\frac{\mathcal{P}_s+\varepsilon\mathcal{P}_a}{(\mathcal{H}_s+\mathcal{P}_s)+\varepsilon(\mathcal{H}_a+\mathcal{P}_a)}\right)\mathcal{H}_s \\
\frac{d\mathcal{P}_s}{dt} &= -V\mathcal{P}_s+T\rho\left(\frac{\mathcal{P}_s+\varepsilon\mathcal{P}_a}{(\mathcal{H}_s+\mathcal{P}_s)+\varepsilon(\mathcal{H}_a+\mathcal{P}_a)}\right)\mathcal{H}_s \\
\frac{d\mathcal{H}_a}{dt} &= \lambda g\mathcal{H}_a\left(1-\frac{(\mathcal{H}_s+\mathcal{H}_a)}{k}\right)-T\left(\frac{\mathcal{P}_a+\varepsilon\rho\mathcal{P}_s}{(\mathcal{H}_a+\mathcal{P}_a)+\varepsilon(\mathcal{H}_s+\mathcal{P}_s)}\right)\mathcal{H}_a \\
\frac{d\mathcal{P}_a}{dt} &= -V\mathcal{P}_a+T\left(\frac{\mathcal{P}_a+\varepsilon\rho\mathcal{P}_s}{(h\mathcal{H}a+\mathcal{P}_a)+\varepsilon(\mathcal{H}_s+\mathcal{P}_s)}\right)\mathcal{H}_a.
\end{aligned} \tag{9}$$

Using these equations, the fixed points for purely sexual and purely asexual populations can be found:

$$(\mathcal{H}_s^*,\mathcal{P}_s^*)=k\left[\frac{V+g-T\rho}{g}\right]\left(1,\frac{T\rho-V}{V}\right)$$

and

$$(\mathcal{H}_a^*,\mathcal{P}_a^*)=k\left[\frac{V+\lambda g-T}{\lambda g}\right]\left(1,\frac{T-V}{V}\right).$$

Stability analysis shows that the purely sexual population is always unstable to invasion by asexuals, and that the asexual population is also unstable if $\lambda\rho<1$. (A more detailed derivation can be found in Keeling & Rand, 1995).

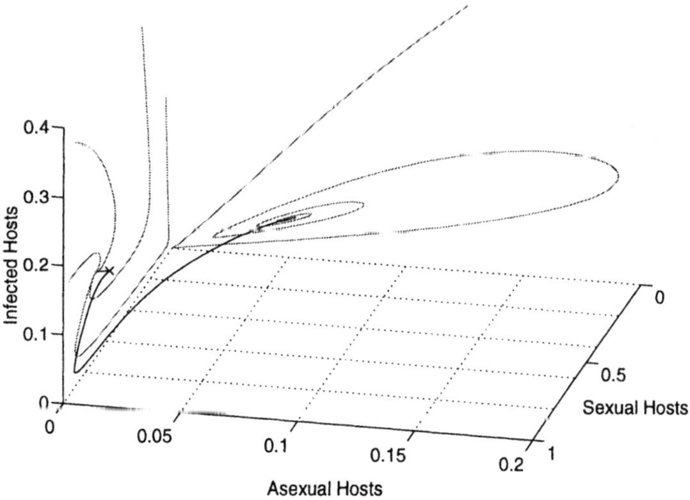

Figure 18. Graph showing the sexual only and asexual only orbits (grey) and perturbation from these to the non-trivial fixed point (black). $V=0.8$, $g=0.5$, $T=1.7$, $\rho=0.588$, $\lambda=2$ and $\varepsilon=0.1$

The position of the non-trivial four-dimensional fixed point

$$(\mathcal{H}_s^*,\mathcal{P}_s^*,\mathcal{H}_a^*,\mathcal{P}_a^*)$$

cannot be found analytically, so to simplify matters the case $0 < \varepsilon \ll 0$ (high aggregation, low coupling) will be examined. It should first be noted that if $\varepsilon = 0$ then the non-trivial fixed point is unattainable and one of the populations decays to zero: the sexual population will dominate if

$$T\rho + g\lambda < T + g \tag{10}$$

This type of behaviour is shown in Figure 18. Both the sexual and the asexual populations are unstable. However, the stable fixed point lies very close to the purely sexual fixed point, so it could be assumed that in a stochastic system the asexuals would be highly prone to extinction.

4.5. CONCLUSION

This section has shown four very different extensions to the simple model addressed in section 3. Each extension has exhibited its own characteristic behaviour and associated problems and difficulties. Allowing more than one host and one parasite shows the potential for co-existence and complex, chaotic dynamics as well as demonstrating that spatial correlations and fluctuations at the local scale are more important to the evolution of short lived parasites than long lived hosts. Increasing the range of both host and parasite has shown a natural way to interpolate between the usual Von Neumann PCA and the standard mean-field approach; but should some more complex neighbourhood structure be used so that the amount of interaction decreases with distance, one should be able to obtain a transition between PCA and reaction-diffusion systems. The work on more realistic evolution is a plausible step in the right direction and it demonstrates how small scale physiological factors can effect long term evolutionary trends. The final part exhibits a method for persistence of sexual reproduction in a population which doesn't rely on large oscillations in the host and parasite genotype but depends on the local diversity of a sexual population; this spatial model also predicts that a sexual population can be stable to slow evolution towards asexuality. The differences between these four models and their mean-field counterparts can be explained in terms of spatial structure and hence correlations or the relatively large fluctuations observed by parasites in their environment.

5. Correlation models and fluctuating systems

All the differences between the probabilistic spatial models and the mean field theory seen so far in this chapter can be accounted for by spatial correlations or stochastic behaviour. The idea that aggregation could affect population dynamics has been profitably used in both ecological and epidemiological models (Hassell & May, 1974; Anderson & May, 1992). However, other than examples of simple birth-death processes from Markov chain theory, stochasticity has been largely over-looked in modelling despite its importance at low numbers. In this chapter methods will be developed that model these two phenomena implicitly. This has the advantage that the results are repeatable and in general computationally inexpensive. The first section will be concerned with spatial correlations, using pairwise interactions, and the second section will deal with the effects of stochasticity and show how the presence of local randomness can effect the global dynamics.

5.1. SPATIAL CORRELATIONS

In previous work spatial correlations have mainly been used to model aggregation of particular individuals. However, we wish to be able to include and predict all manner of local correlations. We shall begin by describing the simple pairwise approach to modelling the correlations before extending this to encompass more complex situations.

5.1.1. *Pairwise Analysis*

The standard mean field equations that are used as the usual approximation to the dynamics of a system assume that the states of a cell's neighbours are independent of the state of the cell. The earlier part of this chapter is a catalogue of situations where this approximation breaks down, often with very profound results. As an improved approximation to modelling the spatial structure, pairs of sites can be considered but with mean-field equations being applied to these pairs (Matsuda *et al.*, 1992; Satō *et al.*, 1994; Altmann, 1995). Let Ω be the set of all possible states for any one cell. The following probabilities can now be defined:

$$
\begin{aligned}
S_A &= \mathbb{P}(\text{cell is of type } \mathbf{A}), \\
P_{AB} &= \mathbb{P}(\text{finding an } \mathbf{AB} \text{ pair}), \\
D_{AB} &= P_{AB}/S_A S_B,
\end{aligned}
$$

where $A, B \in \Omega$. So S_A is the probability of finding the state **A** at a single randomly chosen site and D_{AB} is the modification to the single probabilities (S) necessary due to the correlations between the adjoining double states **A** and **B**. Standard mean-field equations assume no correlations between neighbouring cells, hence $D_{AB} = 1$; the pairwise approximation allows D_{AB} to vary but assumes that there is no correlation between the neighbouring pairs.

Figure 19 shows the results obtained by making this pairwise approximation for the simple host parasite model of section 3; this should be compared to Figure 3.

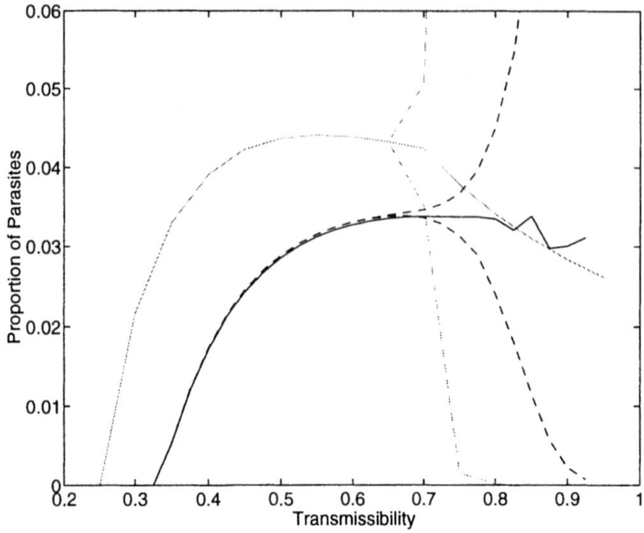

Figure 19. The long-term time-average proportion of parasites for a range of transmissibilities; results from the pair approximation are in black, results from standard mean field theory are in grey, with maximum and minimum values as dashed lines.

The iterations of the pairwise approximation are performed in the following way. As the cellular automaton uses the Von Neumann type neighbourhood each member of a pair is assumed to be in contact with three other neighbours. The probability of these neighbours being in a given state is calculated using the single probability and double correlations given above. For example the probability of a pair of sites being in states **A** and **B** is:

$$\mathbb{P} = S_A S_B D_{AB}$$

Now the probability that a given neighbour of **A** is in state ω is:

$$\mathbb{P} = S_\omega D_{A\omega}$$

In this way the probabilities for each possible neighbourhood configuration can be calculated, and from this the behaviour of each element of the **AB** pair can be found.

The pairwise analysis is in closer agreement with the cellular automaton data in two main respects; the first is the higher value of T_{min}, the lowest transmissibility at which the parasites can survive. T_{min} as been increased from $\frac{1}{4}$ to $\frac{1}{3}$ as each parasitise must have come from one of its surrounding neighbours and so can only infect at most three others. The second advantage is the over-all lower parasite numbers predicted by the pairwise analysis. However, the inclusion of pairwise correlations has lead to a stabilising effect on the dynamics so that no critical transmissibility is observed.

This is due to strong levels of aggregation between the host sites which can be quickly exploited by any parasites finding the hosts. Obviously some factor is being over-looked, this is likely to be the fact that there is some correlation between pairs in the cellular automaton and the next step would therefore be to consider triples. However this process could be carried on repeatedly and would not lead to a greater understanding of the problem. Instead a predetermined correlation between pairs will be used.

5.1.2. *Extended Pairwise Analysis*

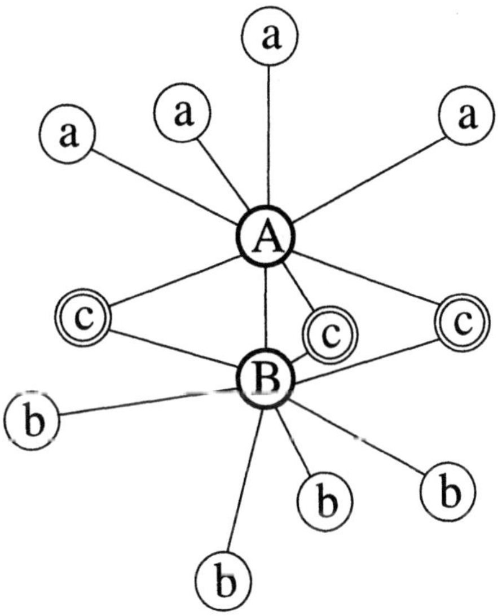

Figure 20. A schematic illustration showing the configuration of an idealised spatial neighbourhood.

Figure 20 shows a example of a typical pair **A**,**B** together with their associated neighbours **a**,**b** and **c**. The neighbours of **A** only are labelled **a**, the neighbours of **B** only are labelled **b** and the neighbours common to both are labelled **c**. It should be noted that in this model the neighbourhoods are not limited to only four cells and the number of neighbours is not even a constant. Instead, there is a probability assigned to each configuration of the neighbourhood. Similar to the way the double modification D_{AB} was defined the triple modification T_{ABC} is given by:

$$T_{ABC} = \frac{\mathbb{P}(\text{finding an } \mathbf{ABC} \text{ triple})}{D_{AB}D_{BC}S_A S_B S_C}$$

where **A**, **C** $\in \Omega$ are the states of two neighbours of a site which is in state **B** $\in \Omega$. For simplicity of notation and from looking at results from various cellular automaton simulations, the following approximation can be made:

$$T_{ABC} = \begin{cases} T_{AB} & \text{if } A = C \\ 1 & \text{otherwise} \end{cases}$$

This substantially reduces the number of T parameters that have to be found, from the number of states cubed to the number of states squared. These T are fixed values either to be measured from the actual system in question or to be approximated from observations. If $T_{\omega A} \equiv 1$ then the pairs are all considered to be independent and the simple pairwise analysis method is regained.

Iterations are performed in a similar manner to the simple pairwise approximations: the probability of a given pair together with a specific neighbourhood type is calculated taking into account the triple correlations ($T_{\omega A}$) as well as the single (S_A) and double (D_{AB}) values. It should be noted that as the triple correlations are fixed parameters it is necessary to normalise the probabilities so that they sum to one. Once the chance of any given pair and neighbourhood configuration has been calculated the probabilistic behaviour of the central pair of sites can be determined from the local rules. The new density of pairs allows the variables S and D to be calculated, and these are then used in successive iterations of the model.

5.1.3. *Simple Host-Parasite System*

To see how this new extended method compares with the original pair analysis we return to the simple host parasite system. Due to the Von Neumann type neighbourhood there are no common neighbours of type **c**; this greatly simplifies the calculation. (Keeling *et al.* (1997b) take advantage of the more complex neighbourhood types that are possible with this method to explain the spread and persistence of measles in a structured population.) For the host-pathogen model,

$$\Omega = \{E, H, P\}$$

	$T_{\omega A}$	E (ω)	H (ω)	P (ω)
A	E	1.0	1.0	1.4
A	H	1.2	1.0	1.4
A	P	0.9	1.0	1.3

The values of the triple correlations were obtained from the cellular automaton with transmissibility $T = 0.5$. The values for other transmissibilities did not differ widely but as the critical transmissibility T_c was approached obtaining results became increasingly difficult because of the frequent extinction of the parasite.

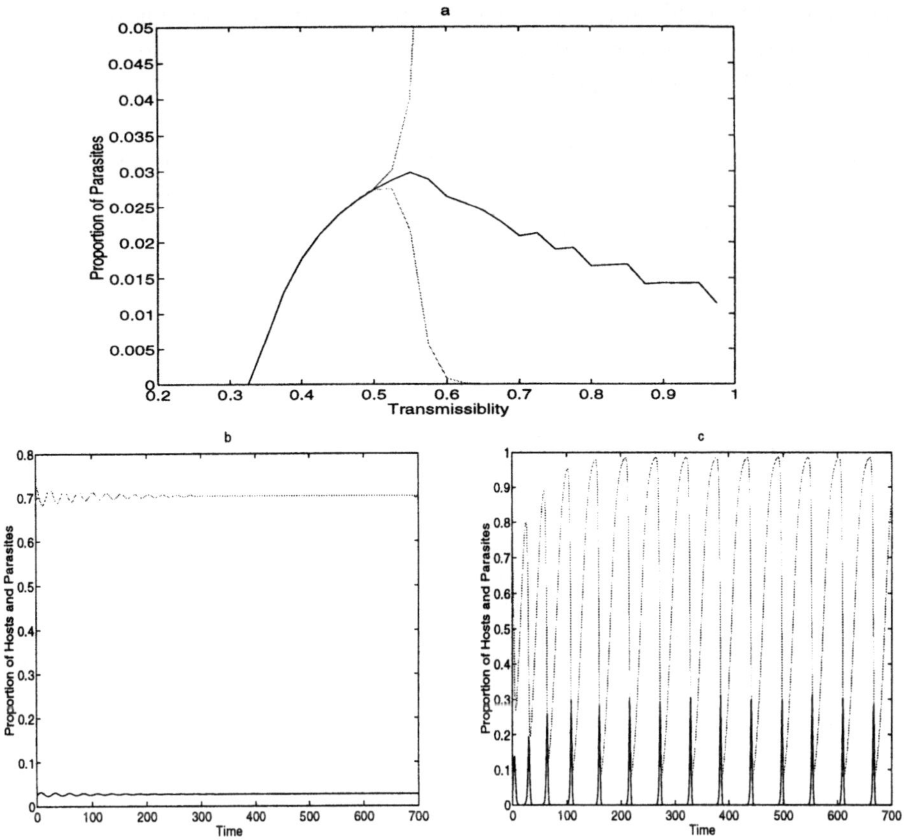

Figure 21. Graph a shows the maximum, minimum (grey) and average (black) proportion of parasites against various transmissibilities. The data is taken from 500 iterations after 500 iterations have been ignored to remove transience. Graphs b and c are the proportions of hosts (grey) and parasites (black) from simulations with transmissibilities $T = 0.5$ and $T = 0.8$ respectively.

The results of the extended pairwise analysis are shown in Figure 21. Graph a shows the maximum, minimum and average number of parasites for various levels of transmissibility. When the effect of the triple correlations is added the onset of instabilities is far sooner and from the minimum value it can be seen that on a grid of 100×100 extinction should be expected around $T = 0.625$. This matches Figure 3 more closely than the results from the uncorrelated pairwise analysis and also gives a good approximation to T_c. Graphs b and c show examples of the temporal behaviour at transmissibilities $T = 0.5$ and $T = 0.8$ respectively. When $T = 0.5$ fast convergence to a fixed equilibrium is seen. However, at around $T = 0.55$ a bifurcation occurs with the equilibrium becoming unstable. For transmissibilities larger than 0.625 the minimum value reached during oscillations has reduced to such an extent that survival of the parasites is unlikely. From graph c it is clear that in a finite model the parasites could not recover from the low troughs.

It appears that the addition of correlations between the pairs in this method increases its accuracy and leads to greater qualitative agreement. However, this greater accuracy is only gained at the expense of some of the speed and simplicity of the original method. Also, the measurement of the triple correlation T_{AB} from a real ecosystem would be very difficult and the results are likely to be far less strong due to the larger neighbourhood and longer ranges of contact.

5.2. LARGE SCALE FLUCTUATIONS

As well as the spatial correlations that probabilistic cellular automata introduce, the fact that the system is stochastic and individual based also has a profound effect on the dynamics. In this section a simple method is introduced for incorporating the large fluctuations seen at a local scale. This method of large fluctuations is initially defined for a general PCA, and then parameters are found and explicit formulae given for the simple host-parasite system of section 3 so that a quantitative evaluation can be made.

5.2.1. *General Definitions*

First some basic but fairly abstract notation needs to be defined to simplify the formulation of the model, this notation made more exact when the simple host-parasite model is discussed. One of the main controlling parameters of the method is N, the number of cells over which the large scale fluctuations occur. N is usually taken as the number of cells in the neighbourhood (including the centre cell); if N is increased the relative size of the fluctuations is reduced. Let Ω be the set of possible states for each cell and let the number of states be D. The following two sets can now be defined:

$$\begin{aligned} \mathcal{T} &= \left\{ \underline{x} \in [0,1]^D : \sum_{\omega \in \Omega} x_\omega = 1 \right\}, \text{ and} \\ \mathcal{S}_N &= \left\{ \underline{x} \in \mathcal{T} : N\underline{x} \in \mathbb{N}^D \right\}. \end{aligned}$$

$\mathcal{T}$ is a simplex formed by the set of all possible proportions of the species and is the space on which the homogeneous mean-field dynamics act. $\mathcal{S}_N$ is the set of all possible proportions of the species found in the configurations of an N cell neighbourhood. Let $\underline{f} : \mathcal{T}^2 \to \mathcal{T}$ be such that $\underline{f}(\underline{t}, \underline{\tau})$ gives the expected result from a population of type $\underline{t}$ which has been acted on by a population $\underline{\tau}$. The standard mean-field theory gives the forward iterations as

$$\underline{t}' = \underline{f}(\underline{t}, \underline{t}). \tag{11}$$

To incorporate the stochastic fluctuations it is necessary to investigate the behaviour of each of the possible N cell neighbourhoods given by $\mathcal{S}_N$. Let Δ be the space of all probability distributions δ on $\mathcal{S}_N$. $\delta(\underline{s})$ is the probability that a randomly chosen N cell neighbourhood has states in the proportions given by $\underline{s}$. It is on the space Δ that the iterations of the fluctuations map will be defined.

The stochasticity can be modelled in terms of a multinomial function M. This has the additional feature that it can be used to take elements of $\mathcal{T}$ to a discretised set,

$$M : \mathcal{T} \to \Delta \qquad \text{with} \qquad M(\underline{t})(\underline{s}) = N! \prod_{\omega \in \Omega} \frac{t_\omega^{(Ns_\omega)}}{(Ns_\omega)!}.$$

Simply considering the behaviour of the N cell neighbourhoods in isolation the following iteration scheme is obtained:

$$\delta' = \sum_{\underline{s} \in \mathcal{S}_N} \delta(\underline{s}) M\left(\underline{f}(\underline{s}, \underline{s})\right).$$

This takes each neighbourhood in $\mathcal{S}$, which exists with density δ, applies an iteration of the mean-field equations and then redistributes the results using the multinomial function. However, with this over-simplified system there is no escape from the trivial unstable fixed points of the mean field equations. This problem arises because the effect of cells external to the neighbourhood has been ignored. A more realistic iteration scheme is obtained using a coupling μ between the discretised neighbourhood and the mean value,

$$\delta' = \sum_{\underline{s} \in \mathcal{S}_N} \delta(\underline{s}) M\left((1-\mu)\underline{f}(\underline{s}, \underline{s}) + \mu \underline{f}(\underline{s}, \widehat{\underline{s}})\right), \tag{12}$$

where $\widehat{\underline{s}}$ is the global mean value,

$$\widehat{\underline{s}} = \sum_{\underline{s} \in \mathcal{S}_N} \delta(\underline{s}) \underline{s}.$$

If it is assumed that there are no correlations between the neighbourhood and the surrounding cells then μ can be taken as μ_0, the ratio between external cells that affect the neighbourhood and the total number of cells that affect the neighbourhood, including all multiplicities. For the Von Neumann neighbourhood $N = 5$ and $\mu_0 = \frac{12}{20}$ (there are five cells each of which is affected by four surrounding sites, giving 20 cells in total; and there are four cells each of which have three external neighbours giving 12), and for the larger Moore neighbourhood (including the "diagonal" neighbours) $N = 9$, $\mu_0 = \frac{4}{9}$. The value of μ can be further modified by considering the local correlations, if the correlations are high then the cells of the neighbourhood are more likely to be affected by cells of a similar type hence μ should be reduced. μ can be assumed a decreasing function of the local correlations and should be equal to zero only when the correlations are maximal.

As $N \to \infty$ the set $\mathcal{S}_N$ becomes more dense until finally every point in $\mathcal{T}$ is infinitesimally close to a point in $\mathcal{S}_N$. Also as N increases the multinomial function M becomes tighter until it maps elements of $\mathcal{T}$ to near point distributions in Δ. Increasing N also decreases μ_0 as seen in the above example of the 4 cell Von Neumann and the 9 cell Moore neighbourhoods. Hence if the mean field theory has a globally attracting fixed point then as N becomes large the fluctuation method (equation (12)) tends to the mean field approximation (equation (11)).

Although this method is primarily designed to model probabilistic cellular automata it can easily be adapted to predict the behaviour of other spatial systems. The fluctuation method assumes that the stochasticity is introduced by specifying the probability that a discrete change will occur in an iteration; this leads to the multinomial distributions. In some systems however, such as integer based coupled map lattices, events occur with a Poisson distribution; the function M can be simply modified to take this into account.

5.2.2. *Finding μ for a Specific Problem*

In the following sections the fluctuation method is applied to the simple host-parasite system to illustrate three methods of finding μ. In this simple system $\Omega = \{E, H, P\}$ corresponding to Empty, Host and Parasitised cells, and hence $\mathcal{T} = \{(E, H, P) : E+H+P = 1\}$. Because the cellular automaton uses the four cell Von Neuman neighbourhood, N is taken to be 5. The abstract definitions of the previous section can now be made more explicit. The iteration function is

$$\underline{f}\left(\begin{pmatrix} e \\ h \\ p \end{pmatrix}, \begin{pmatrix} E \\ H \\ P \end{pmatrix}\right) = \begin{pmatrix} e + pV - e\left[1-(1-gH)^4\right] \\ h + e\left[1-(1-gH)^4\right] - h\left[1-(1-TP)^4\right] \\ p(1-V) + h\left[1-(1-TP)^4\right] \end{pmatrix}$$

(*cf.* equations (4)), and the multinomial funciton is

$$M\begin{pmatrix} E \\ H \\ P \end{pmatrix}\begin{pmatrix} e \\ h \\ p \end{pmatrix} = 5!\frac{E^{5e}H^{5h}P^{5p}}{(5e)!(5h)!(5p)!}.$$

Trial and Error Figure 22 graphs *a* and *b* show the proportion of parasites and hosts respectively predicted for the simple host-parasite system by equation (12), for a range of μ. As can be seen from the graphs, both the proportion of parasites and the proportion of hosts compare favourably with the results from the cellular automaton at $\mu \approx 0.23$. This value of μ is a lot smaller than the theoretical uncorrelated value of $\mu_0 = 0.6$, showing strong positive correlations.

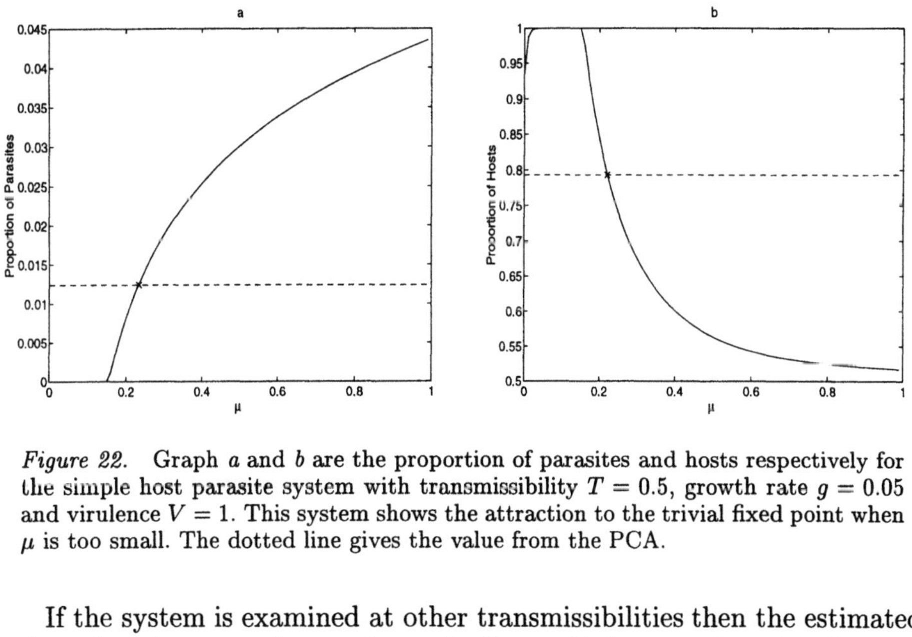

Figure 22. Graph *a* and *b* are the proportion of parasites and hosts respectively for the simple host parasite system with transmissibility $T = 0.5$, growth rate $g = 0.05$ and virulence $V = 1$. This system shows the attraction to the trivial fixed point when μ is too small. The dotted line gives the value from the PCA.

If the system is examined at other transmissibilities then the estimated value of μ does not change dramatically until close to the critical transmissibility. As the critical transmissibility is approached the correlations in the system become stronger as each parasite is more likely to infect the surrounding cells. This leads to a decrease in μ which in turn leads to an increasing tendency to be attracted to the trivial fixed point, i.e. for the parasites to die out.

Ignoring the fact that μ decreases as the critical transmissibility is approached, Figure 23 graph *a* shows the proportion of parasites for various transmissibilities using $\mu = 0.23$. This graph should be compared to Figure 3. The lower transmissibility threshold has clearly been captured and the overall behaviour, especially the lower number of parasites, is comparable to the proportion of parasites shown as the survived average.

Graph *b* shows the proportion of parasites, keeping $\mu = 0.23$, $T = 0.5$, and varying the neighbourhood size N. This can be compared to the results in section 4. As N is increased it is clear that the fluctuation method approaches the values from the mean field approximation. This graph is obtained by keeping the value of μ constant. However, as the neighbourhood over which the interactions take place increases then the theoretical uncorrelated μ_0 will increase. Also, increasing N will lead to a decrease in the correlations between sites which will cause a further increase in μ. These effects will cause the actual system to approach the mean field approximations faster than predicted in graph *b*, which is more in keeping with results from the PCA.

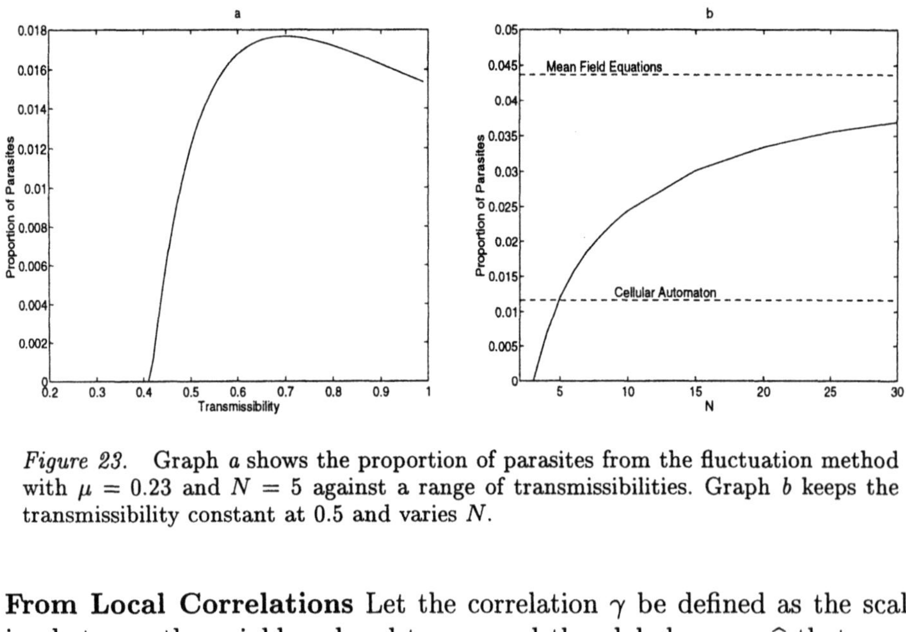

Figure 23. Graph *a* shows the proportion of parasites from the fluctuation method with $\mu = 0.23$ and $N = 5$ against a range of transmissibilities. Graph *b* keeps the transmissibility constant at 0.5 and varies N.

From Local Correlations Let the correlation γ be defined as the scaling between the neighbourhood type $\underline{s}$ and the global means $\widehat{\underline{s}}$ that most closely approximates the type of sites $\underline{t}(\underline{s})$ surrounding the neighbourhood in question (see Figure 24). This means the γ is chosen to minimise

$$\sum_{\underline{s}} \delta(\underline{s}) \left\| \underline{t}(\underline{s}) - (\gamma \underline{s} + (1-\gamma)\widehat{\underline{s}}) \right\|^2$$

$\delta(\underline{s})$, $\underline{t}(\underline{s})$ and $\widehat{\underline{s}}$ can all be measured from the PCA giving $\gamma_1 = 0.721$ and $\mu = \mu_0(1-\gamma_1) = 0.16$ which from figure 22 appears to be very low. A better approximation can be obtained if γ is chosen so that it most closely approximates the effects of $\underline{t}$ on $\underline{s}$, i.e. it minimises

$$\sum_{\underline{s}} \delta(\underline{s}) \,\| f(\underline{s}, \underline{t}) - f(\underline{s}, \gamma \underline{s} + (1-\gamma)\hat{\underline{s}}) \|^2$$

From the PCA this leads to $\gamma_2 = 0.532$ and $\mu = 0.28$. This is clearly a far more accurate result.

From Within the Model Although the last two methods of obtaining a value of μ gave fairly accurate results, they rely on the use of the cellular automaton. Having to know the answer before the parameters can be determined is not particularly desirable even though the parameters obtained could be used to look at other related problems. What is required is a method of finding μ which is independent of the computationally expensive cellular automaton, i.e. a method which can calculate μ from within the fluctuation equations. This would also have the advantage that it could cope with a temporally varying μ if the model did not tend to a statistically stationary attractor but had either oscillatory or chaotic dynamics.

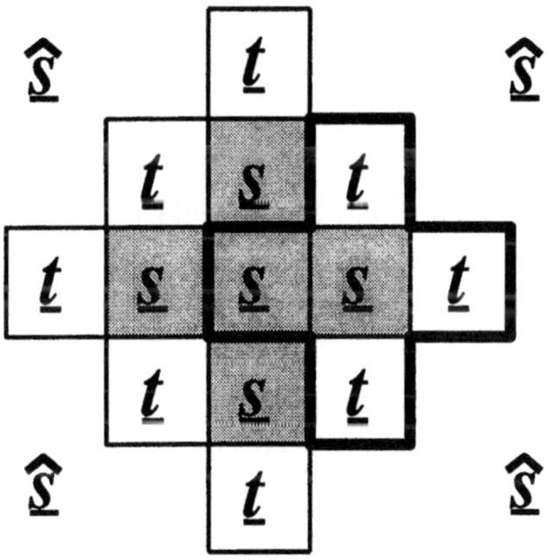

Figure 24. The layout of a five cell configuration $\underline{s}$ together with its nearest neighbours $\underline{t}$.

Parameterising from within the model can be accomplished by considering a five site configuration as highlighted on the right-hand side of Figure 24. This is composed of two sites of type $\underline{s}$ and three of type $\underline{t}$; however, δ gives us the distribution of all five site configurations so the average $\underline{t}(\underline{s})$ can be approximated. The calculation of γ and hence μ can proceed as before, choosing γ to approximate the effects of $\underline{t}$ gives the best results again.

Figure 25 shows the values of γ_2 and μ and the corresponding proportion of parasites. Despite the lower correlations present in this system compared to the previous two, there still exists a lower threshold transmissibility that compares well to the cellular automaton model and there is also a upper

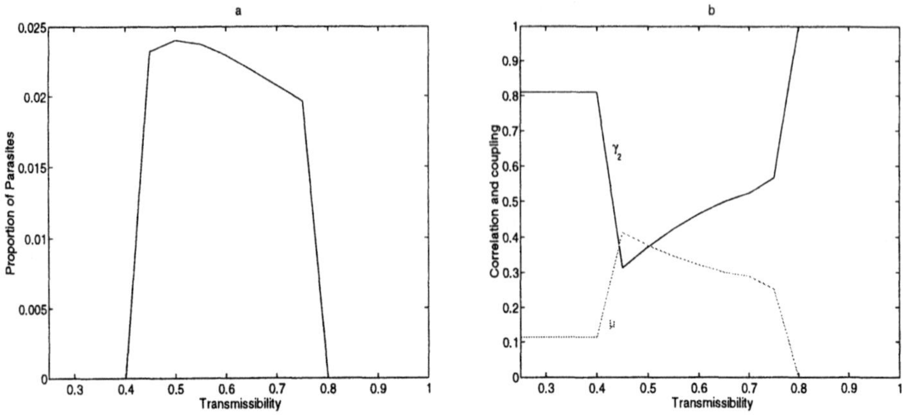

Figure 25. Graphs a and b are for a range of transmissibilities for the simple host parasite model. Graph a shows the proportion of parasites predicted by the fluctuation method. Graph b shows the value of γ_2 and the corresponding μ that are calculated from within the model.

critical transmissibility but this is higher than the one found in the cellular automaton.

The main inaccuracies in this method are the assumption that μ is a scalar variable, for in truth the correlations should be given by a matrix. However this would be difficult to interpret and some values would be difficult to obtain due to the low density of many neighbourhood types. Another inaccuracy is the fact that f does not take into account the finite structure of the neighbourhood. The final problem is that the multinomial function M does not give any consideration to the local pairwise correlations that can exist between neighbouring pairs and often produces a configuration in which all five cells are parasitised, which is impossible in the spatial model.

5.3. CONCLUSION

In this chapter two main methods have been outlined the first implicitly models the spatial correlations and the second also includes the effects of local stochasticity. Both these features affect the dynamics of the system, leading in general to lower levels of parasitism and more closely matching the cellular automaton simulations. Both these models have the advantage of deterministic results so that there is no need for the vast number of averages usually taken with most probabilistic simulations. The other main advantage of these methods is that they rely on extensions to the simple mean-field theory equations and hence can be applied to almost any modelling situation. Unfortunately the complexity of the models grows rapidly with the number of species and the number of interacting cells. However, it is still hoped that these techniques will give some insights into the roles of space and stochasticity in real world ecological and epidemiological situations.

6. Conclusions

All of the above models have demonstrated the importance of space and discrete populations when studying the dynamics of diseases and parasites. Although each model has its own complications and peculiar characteristics the differences between the PCA and the mean-field dynamics can be explained by spatial correlations or stochastic fluctuations. It should be noted that all of the models described here are caricatures of the real systems and hence many of the effects observed in this work are greatly exaggerated. Very few ecosystems exhibit 'nearest neighbour' interactions and a larger neighbourhood has been shown to decrease correlations and stochastic fluctuations, hence leading to results closer to mean-field dynamics. To reduce computation time and system sizes the local rules have been made very extreme: parasites have been given severely detrimental effects and hosts have relatively short generation times. With smoother, slower rules the stochastic fluctuations are likely to be less frequent and of smaller magnitude.

All the models have only dealt with spatial structure. Superimposed on this is social structure and in particular age-structure, which is known to play a very important role in the transmission of many diseases. The model of sexual reproduction gives an insight into the inclusion of genetic heterogeneity, but complex non-linearities in translating genotype to phenotype together with spatial correlations could produce all manner of dynamics from punctuated evolution to speciation. Almost all spatial models presume that there are no underlying heterogeneities in the space; in the natural world this simplification is rarely the case and variations abound. Gradients in the transmissibility, or regions of high host density, are realistic scenarios that may have interesting effects.

There appears to be a vast amount of work to be done in extending spatial models in a variety of directions. However, for every new feature that is included an array of new parameters need to be found. Therefore as much emphasis also needs to be placed on new methods of data capture and parameter estimation to fuel the growing numbers of spatial models. Finally greater progress might be made in the long term if we were to concentrate on capturing and predicting the essential differences between homogeneous and spatial systems, rather than on devising ever more computationally expensive simulations. This is the issue addressed in section

5 and other such ways of extending the mean-field equations may prove invaluable in our attempt to understand, predict, and finally control the spread of a disease through the population.

References

Altmann, M. (1995) Susceptible-Infectious-Recovered epidemic models with dynamics partnerships *J. Math. Biol.* **33**, pp. 661–675.

Anderson, R.M. & May, R.M. (1992) *Infectious diseases of humans.* Oxford University Press, Oxford.

Barkham, J.P. & Hance, C.A. (1982) Population Dynamics of the Wild Daffodil, *J. of Ecology* **70**, pp. 323-344.

Begon, M., Harper, J.L. & Townsend, C.R. (1990) *Ecology. Individuals, Populations and Communities.* Blackwell Scientific Publications, Oxford.

Bell, G. (1982) *The masterpiece of nature: the evolution and genetics of sexuality.* University of California Press.

Bollobás, B. (1985) *Random Graphs.* Academic Press.

Bremermann, H.J. (1980) Sex and polymorphism as strategies in host-pathogen interactions, *J. Theo. Biol.* **87**, pp. 641-702.

Brown, W. (1993) Cheat thy Neighbour, a recipe for success *New Scientist* **1894**, p. 19.

Darwin, C. (1862) *J. Proc. Linn. Soc. (Botany)* **6**, pp. 77–96.

Durrett, R. & Levin, S.A. (1994a) Stochastic Spatial Models: A User's Guide to Ecological Applications, *Phil. Trans. R. Soc. Lond.* **343**, pp. 329-350.

Durrett, R. & Levin, S.A. (1994b) The Importance of Being Discrete and Spatial, *Theo. Pop. Biol.* **46**, pp. 363-394.

Ermentrout, G.B. & Edelstein-Keshet, L. (1992) Cellular Automata Approaches to Biological Modelling, *J. Theo. Biol.* **160**, pp. 97-133.

Glesener, R.R. & Tilman, D. (1978) Sexuality and the components of environmental uncertainty: Clues from geographic parthenogenesis in terrestrial animals, *Am. Nat.* **112**, pp. 659-673.

Hamilton, W.D. (1980) Sex versus non-sex versus parasite, *Oikos* **35**, pp. 282-290.

Hamilton, W.D., Axelrod, R. & Tanese, R. (1990) Sexual reproduction as an adaptation to resist parasites (A Review), *Proc. Natl. Acad. Sci.* **87**, pp. 3566-3573.

Hassell, M.P., Comins, H. & May, R.M. (1991) Spatial Structure and Chaos in Insect Population Dynamics, *Nature* **353**, pp. 255-258.

Hassell, M.P. & May, R.M. (1974) Aggregation in predators and insect parasites and its effects on stability, *J. Anim. Ecol.* **43**, pp. 567-594.

Keeling, M.J., Mezic, I., Hendry, R.J., McGlade, J. & Rand D.A. (1997a) Characteristic length scales of spatial models in ecology, *Phil. Trans. Roy. Soc. Lond. B* **352**, pp. 1589–1601.

Keeling, M.J. & Rand, D.A. (1995) A spatial mechanism for the evolution and maintenance of sexual reproduction, *Oikos* **74**, pp. 414–424.

Keeling, M.J, Rand, D.A., & Morris, A.J. (1997b) Correlation models for childhood epidemics *Proc. Roy. Soc. Lond. B* **264**, pp. 1149–1156.

Levin, D. (1975) Pest pressure and recombination systems in plants, *Am. Nat.* **109**, pp. 437-451.

Lively, C.M. & Howard, R.S. (1994) Selection by parasites for clonal diversity and mixed mating, *Phil. Trans. R. Soc. Lond.* **346**, pp. 271-281.

Matsuda, H., Ogita, N., Sasaki, A. & Satō, K. (1992) Statistical mechanics of population: The lattice Lotka-Volterra model, *Prog. Theor. Phys.* **88**, pp. 1035-1049.

Maynard Smith, J. (1976) *The evolution of sex.* Cambridge University Press, Cambridge.

Mollison, D. (1977) Spatial Contact Models for Ecological and Epidemic Spread, *J. R.*

Statistical Soc. B **39**, pp. 283-326.
Mollison, D. (1986) Modelling Biological Invasion: chance, explanation, prediction, *Phil. Trans. R. Soc. Lond.* B **314**, pp. 675-693.
Mollison, D., Isham, V. & Grenfell, B. (1993) Epidemics : Models and Data, *J. Roy. Statistical Soc.* A **157**, pp. 115-149.
Nicholson, A.J. & Bailey, V.A. (1935) The Balance of Animal Populations, Part I, *Proc. Zool. Soc. Lond.* **3**, pp. 551-598.
Pimm, S.L. (1991) *The Balance of Nature?* University of Chicago Press. pp. 173-188.
Pool, R. (1989) Is it Chaos or is it just Noise? *Science* **243**, pp. 25-28.
Rand, D.A., Keeling, M.J. & Wilson, H.B. (1995) Invasion, stability and evolution to criticality in spatially extended, artificial host-pathogen ecologies, *Proc. R. Soc. Lond.* B **259**, pp. 55-63.
Satō, K., Matsuda, H. & Sasaki, A. (1994) Pathogen invasion and host extinction in lattice structured populations, *J. Math. Biol.* **32**, pp. 251-268.
Schrag, S.J., Mooers, A.O, Ndifon, G.T. & Read, A.F. (1994) Ecological correlates of male outcrossing ability in a simultaneous hermaphrodite snail, *Am. Nat.* **143**, pp. 636-655.
Sugihara, G., Grenfell, B.T. & May, R.M. (1990) Distinguishing error from chaos in ecological time series, *Phil. Trans. R. Soc. Lond. B* **330**, pp. 235-251.
Van de Bosch, F., Metz, J.A.J. & Diekmann, O. (1990) The Velocity of Spatial Population Expansion, *J. Math. Biol.* **28**, pp. 529-565.
Vrijenhoek, R.C. (1993) The origin and evolution of clones versus the maintenance of sex in Poeciliopsis, *J. Heredity* **84**, pp. 388-395.
Wilson, E.O. (1992) *The Diveristy of Life*. Penguin Science.
Wilson, W.G., De Roos, A.M. & McCauley, E. (1993) Spatial Instabilities within the Diffusive Lotka-Volterra System: Individual-Based Simulation Results, *Theo. Pop. Biol.* **43**, pp. 91-127.

LATTICE DYNAMICAL SYSTEMS

LEONID A. BUNIMOVICH
assisted by C. Giberti

Center for Dynamical Systems and Nonlinear Studies
Georgia Institute of Technology
Atlanta, GA 30332
U.S.A.

Abstract.
Lattice Dynamical Systems (LDS) form a class of extended systems that is intermediate between partial differential equations and cellular automata. Although introduced rather recently, LDS have already helped to clarify some vague notions and ideas that form the basis of the modern phenomonological theory of spatio-temporal dynamics. We discuss some rigorous results that have been obtained in this area.

1. Introduction

Partial Differential Equations (PDEs) form the traditional models that describe the evolution of spatially extended dynamical systems. However, mathematical studies of PDEs tend to be very complicated and for a very long time all the successes in the area were essentially restricted to finding some simple solutions and the study of their stability. Almost all these solutions are classical ones. Much more recently the theory of deterministic chaos has prompted the extensive search for strange attractors in the models described by PDEs. However, again because of the extreme difficulty of analytic studies of solutions with complex behavior, these efforts essentially reduce to the problem of estimating the dimension of an attractor rather than to the investigation of the solutions that live on it (Temam, 1988).

The problem of finding and describing invariant measures on strange attractors of PDEs is almost completely open. One should mention in this respect the recent series of papers by Bourgain (1994a & b, 1996) and

J. C. Robinson and P. A. Glendinning (eds.),
From Finite to Infinite Dimensional Dynamical Systems, 59–83.

McKean & Vaninski (1994), where invariant measures for some nonlinear evolution PDEs were constructed, opening the possibility of applying the methods of statistical mechanics to the corresponding dynamical systems (Lebowitz *et al.*, 1988). However, the problems of studying ergodic properties (ergodicity, mixing, etc. - see Sinai (1994)) of the dynamical systems generated by PDEs at the moment is, essentially, out of reach.

On the other hand, computer simulations have emerged as an effective and powerful tool to study dynamical systems with complex behavior. In such studies the dynamical systems by necessity get discretized in space as well as in time. This was the major reason for introducing the new models of extended dynamical systems (see, for example, Crutchfield & kaneko (1987) or Kaneko (1993a & b)) that have received the name Coupled Map Lattices (CML). CML, in fact, can be considered as systems of interacting (generalized) oscillators where each oscillator is characterized not only by its internal state (internal coordinates) but also by its position in the physical space.

The most popular and well studied CML model is defined as follows:

$$x_i(n+1) = f(x_i(n)) + \frac{\epsilon}{2}\left(f(x_{i-1}(n)) - 2f(x_i(n)) + f(x_{i+1}(n))\right), \quad (1.1)$$

where $i \in \mathbb{Z}$, $n \in \mathbb{Z}_+$, $x_j(m) \in [0,1]$ and $f : [0,1] \to [0,1]$ is some one-dimensional map, e.g., $f(u) = au(1-u)$. The "physical space" in the CML (1.1) is formed by the one-dimensional lattice of integers $\mathbb{Z}$.

This is the characteristic property of the Lattice Dynamical Systems (LDS) that is reflected in their name. CML form the special class of LDS where the dynamics is split into the composition of the local dynamics (generated by the local map f) and the spatial interactions (as in the CML (1.1)) or vice versa.

The general form of LDS is

$$\mathbf{x}(n+1) = F(\mathbf{x}(n)), \qquad \mathbf{x}(n) = \{x_{\mathbf{i}}(n)\}, \quad n \in \mathbb{Z}_+, \quad \mathbf{i} \in \mathbb{Z}^d, \quad d \geq 1 \quad (1.2)$$

in the case of discrete time, or

$$\frac{d\mathbf{x}}{dt} = F(\mathbf{x}(t)) \qquad \mathbf{x}(t) = \{x_{\mathbf{i}}(t)\}, \quad t \in \mathbb{R}_+, \quad \mathbf{i} \in \mathbb{Z}^d, \quad d \geq 1 \quad (1.3)$$

in the case of the continuous time, where $x_{\mathbf{i}}(n)$ ($x_{\mathbf{i}}(t)$) is a point of some finite-dimensional manifold $M_{\mathbf{i}}$ that is called the local phase space at the point $\mathbf{i}$ of the d-dimensional lattice $\mathbb{Z}^d$. Here $\mathbb{Z}^d$ is the cubic lattice in $\mathbb{R}^d$, i.e., the set of all vectors $\mathbf{i} = (i_1, \ldots, i_d)$ with integer coordinates.

Observe that CML form a special class of LDS. Namely, for a CML F in (1.2) or (1.3) can be represented as $F = G \circ H$ (or $F = H \circ G$), where the operator H describes the independent action of the local dynamics at each site of the lattice, while G describes the spatial interactions.

It is worthwhile to mention that sometimes LDS of type (1.2), (1.3) etc. form reasonable models of real systems. This happens when the dynamics is essentially governed by the evolution and interaction of some localized structural elements of a system, e.g., by Taylor vortices in the Taylor-Couette flow (Lesieur, 1989).

The essential feature of LDS (1.2), (1.3) is that these infinite-dimensional dynamical systems describe the evolution of infinitely many similar objects, or even of identical ones if all local phase spaces $M_{\mathbf{i}}$ ($\mathbf{i} \in \mathbb{Z}^d$) coincide and the equations of motion are translationally invariant.

Such a situation is typical for the models in statistical mechanics that describe the evolution of an infinite ensemble of particles interacting via some (translationally invariant) potential. It is worthwhile to mention that PDEs, in contrast, describe an evolution and interaction of (infinitely many) "modes" that are usually essentially nonequivalent because of the constraints related, for example, with the conservation laws.

Therefore, it is natural to explore this analogy of LDS with the models of statistical mechanics and to try to apply to these systems the thermodynamic formalism that has proven to be an efficient tool for studies of nonextended dynamical systems with strongly chaotic behavior (hyperbolic dynamical systems) - see Simon (1994), Ruelle (1978) and Bowen (1975).

This approach was suggested in Bunimovich & Sinai (1988) where LDS were subjected to rigorous mathematical study for the first time. They are now an active area of mathematical research. The first part of these lectures is devoted to a discussion of the current state of the art in this area, which can be viewed as the first room in the building of the ergodic theory of extended systems.

The goal of these studies is to prove that LDS generated by weak spatial interactions of local dynamical systems (generalized oscillators) with strongly chaotic dynamics have the property of space-time mixing. (This property was introduced in Bunimovich & Sinai (1988) (see Sect. 1 for the exact definition), and called there "space-time chaos." However, I am strongly convinced that this term should be changed to "space-time mixing" in order to avoid confusions in the future.)

Obviously, LDS without spatial interactions between local dynamical systems demonstrate space-time chaos provided that this local dynamics is chaotic. Therefore the goal is to prove that this pre-existent space-time chaos does not disappear under weak spatial interaction, i.e., it is in a sense a stable regime of the motion of such systems.

The other direction of the studies of LDS with weak interactions has as its goal the construction of a variety of nontrivial solutions of such systems.

The idea is first to construct such solutions for the (uninteresting) case when there are no spatial interactions, and then to prolong them (by employing an appropriate Implicit Function Theorem in an appropriate space) to the (interesting) region of weak spatial interactions. These studies are closely related to the concept of the so-called anti-integrable limit (Aubry, 1992), which is discussed in MacKay (1996).

The third actively developed areas in these studies of LDS uses the method, well known in PDE theory, of the separation of variables. The basic idea is that solutions with very simple time-dynamics (e.g., the fixed points) may have very complicated (chaotic) spatial structure. This approach, suggested by M. Rabinovich, was developed by V. Afraimovich and coauthors (Afraimovich *et al.*, 1994 & 1996; Afraimovich & Chow, 1995; Afraimovich & Nekorkin, 1994; Afraimovich & Pesin, 1993a & b) - see also Chow & Shen (1995), Shen (1995), and the review in Chow & Mallet-Paret (1995).

All these three directions of study deal with some boundary situations rather than with the intrinsic ones for the dynamics of LDS. Indeed, the first two deal only with small perturbations of the state without spatial interactions and the third one assumes that the motion in time is very simple or completely absent.

The problem of studying LDS with strong spatial interactions is much more difficult. Researchers do have now a rather good intuition about the evolution of point (nonextended) dynamical systems. For instance, their intuition suggested that weak interactions of chaotic local systems should not destroy space-time chaos. However, intuition about the dynamics of spatially extended systems is very poor. It is related to the lack of the collection of exactly solvable (nontrivial) models and to the lack of general mechanisms (bifurcations) that generate chaos-order transitions in extended dynamical systems.

One of the natural questions is about the mechanisms that cause a complex (chaotic) motion in LDS with simple local dynamics. Obviously, such mechanisms may work only in the region of strong spatial interactions, because these interactions should destroy stable simple local dynamics. In Section 2 a first such mechanism is discussed, which we called the peak-crossing bifurcation. It is natural to expect that there should be many such scenarios of transition to chaos in extended systems. Section 3 deals with the problem of stability of spatial structures in LDS. We show that, at least in some cases, a spatial structure gets destroyed via the loss of stability of its most tiny pattern.

In Section 4 we discuss some results on the relations between chaos in the local systems and the complexity of spatial structures in LDS of the reaction-diffusion type.

2. Space-Time Mixing in Lattices of Chaotic Maps with Weak Spatial Interactions

Let M be a compact Riemannian manifold and

$$\mathcal{M} = \bigotimes_{\mathbf{i} \in Z^d} M_{\mathbf{i}} = \{\mathbf{x} = \{x_{\mathbf{i}}\}, x_{\mathbf{i}} \in M\},$$

where $M_{\mathbf{i}}$ is a copy of M. The sets $M_{\mathbf{i}}$, $\mathbf{i} \in \mathbb{Z}^d$, are called the local phase spaces and $\mathcal{M}$ is called the phase space of the LDS.

We define now the dynamics $\Phi : \mathcal{M} \to \mathcal{M}$ that generates the LDS. Let $f : U \to M$ be a $C^{1+\alpha}$-diffeomorphism (i.e., the derivative of f is Hölder continuous of exponent α) and let $\Lambda \subset U$ be a closed invariant hyperbolic set for f. The last means that the tangent bundle TM can be split into subbundles over Λ, invariant under the differential Df, such that $TM = E^S \oplus E^U$, and for some $0 < \gamma < 1$, $C > 0$,

$$\|Df^n w\| \le C\gamma^n \|w\| \quad \text{for } n \ge 0, \quad w \in E^S;$$

$$\|Df^{-n} w\| \le C\gamma^n \|w\| \quad \text{for } n \ge 0, \quad w \in E^U.$$

In particular, Λ can coincide with M; in this case f is called an Anosov diffeomorphism.

Let $F = \otimes_{\mathbf{i} \in \mathbb{Z}^d} f$ be a mapping $F : \mathcal{M} \to \mathcal{M}$, which is called a lattice of noninteracting maps. One gets a LDS by considering a small perturbation Φ of F.

The group of spatial translations $\{S_{\mathbf{j}}\}$, $\mathbf{j} \in \mathbb{Z}^d$, acts on the phase space of LDS as $(S_{\mathbf{j}}\mathbf{x})_{\mathbf{i}} = x_{\mathbf{i}+\mathbf{j}}$, where $\mathbf{i}, \mathbf{j} \in \mathbb{Z}^d$, $\mathbf{x} = (x_{\mathbf{i}})$, $\mathbf{i} \in \mathbb{Z}^d$.

Therefore the pair (Φ, S) naturally generate space-time dynamics on $\mathcal{M}$. It is easy to see that Φ and S commute.

In order to define the LDS exactly one needs to introduce on $\mathcal{M}$ the structure of a differentiable Banach manifold. To avoid this, and some related long definitions (see Jiang (1995), for example), we will consider in what follows the case when the local phase space M is the unit circle S^1 (Bricmont & Kupiainen, 1995, 1996 & 1997).

In this case the phase space $\mathcal{M} = (S^1)^{\mathbb{Z}^d}$ is the direct product of circles over $\mathbb{Z}^d$. The set $\mathcal{M}$ also can be considered as the collection of maps from the lattice $\mathbb{Z}^d$ to the circle. The product (Tikhonov) topology defines the topology on $\mathcal{M}$ as well as the Borel σ-algebra $\mathcal{B}$.

We will describe now a local dynamical system of LDS. Consider an expanding, orientation preserving $C^{1+\alpha}$, $\alpha > 0$, map $f : S^1 \to S^1$. It is convenient to describe the properties of f in terms of its lift to $\mathbb{R}^1$, which we will denote by $\tilde{f}$. Suppose that (i) $\tilde{f}(0) \in [0, 1]$ and (ii) $\tilde{f}'(x) > \lambda > 1$. Then there exists an integer $k > 1$ such that $\tilde{f}(x + 1) = \tilde{f}(x) + k$ for any $x \in \mathbb{R}^1$, because $\tilde{f}$ is the lift of f.

It is well known that the dynamical systems generated by such maps have very strong chaotic properties (Simon, 1994; Katok & Hasselblatt, 1995). Namely, there exists a unique f-invariant probability measure on S^1 which is absolutely continuous with respect to the Lebesgue measure (these measures form the special class of Sinai-Bowen-Ruelle (SBR) measures). In addition, the following two properties are satisfied:

(a) For any absolutely continuous probability measure ν and for any continuous function g

$$\int g \circ f^n d\nu \underset{n\to\infty}{\longrightarrow} \int g\, d\mu \tag{2.1}$$

This property is sometimes referred to as the relaxation to an equilibrium.

(b) There exist constants $C > \infty$ and $\theta > 0$ such that for any two functions $g \in L^\infty(S^1)$ and $h \in C^\alpha(S^1)$

$$\left|\int g \circ f^{(n)} h\, d\mu - \int g\, d\mu \int h\, d\mu\right| \leq C\|g\|_\infty \|h\|_\alpha e^{-\theta n}, \tag{2.2}$$

where $C^\alpha(S^1)$ is the space of Hölder continuous functions with the norm $\|h\|_\alpha = \|h\|_\infty + \sup\limits_{x,y} \frac{|h(x)-h(y)|}{|x-y|^\alpha}$.

Property (b) is called the exponential decay of correlations. Let $F : \mathcal{M} \to \mathcal{M}$ denote the Cartesian product $F = \bigotimes_{\mathbf{i}\in\mathbb{Z}^d} f_{\mathbf{i}}$, where $f_{\mathbf{i}}$ is a copy of f. We introduce now an operator for the spatial interactions (coupling) $A : \mathcal{M} \to \mathcal{M}$ (Bricmont & Kupiainen, 1997). Consider

$$a_{\mathbf{j}}(\mathbf{x}) = x_{\mathbf{j}} + \epsilon \sum_{\mathbf{k}} g_{|i-k|}(x_j, x_k) \tag{2.3}$$

where g is a $C^{1+\alpha}$ function, periodic in both variables, which satisfies

$$\|g_{|\mathbf{j}-\mathbf{k}|}\| < C_1 \exp(-\kappa\|\mathbf{j}-\mathbf{k}\|),$$

with $C_1 < \infty$ and $\kappa > 0$. Here, $a_{\mathbf{j}}$ gives the lift of $A_{\mathbf{j}}$, and so $A_{\mathbf{j}}$, the projection of A on the $\mathbf{j}^{\text{th}}$ factor, is given by

$$A_{\mathbf{j}} = \exp(2\pi i a_{\mathbf{j}}).$$

Spatial interactions of the analogous form were considered in Bunimovich & Sinai (1988). They belong to the class of weak spatial interactions. A general notion of weak spatial interactions was considered in Pesin & Sinai (1991) (see also Jiang, 1995; Bricmont & Kupiainen, 1996). Roughly speaking the spatial interactions are called weak if their strength decays exponentially with the distance between the sites on a lattice. The coupled map lattice

$\Phi : \mathcal{M} \to \mathcal{M}$ is defined as $\Phi = A \circ F$. A natural invariant measure for CML, first constructed in Bunimovich & Sinai (1988), is the natural generalization of SBR measures to these infinite-dimensional dynamical systems.

A Φ-invariant Borel probability measure μ on $\mathcal{B}$ is a SBR measure if for any finite subset $\Lambda \subset \mathbb{Z}^d$ the restriction μ_Λ of μ to B_Λ is absolutely continuous with respect to m_Λ, where m_Λ is the (product) Lebesgue measure on $\mathcal{M}_\Lambda = (S^1)^\Lambda$.

Theorem 2.1. (Bricmont & Kupiainen, 1997) *Let f and A satisfy the assumptions given above. Then there exists $\epsilon_0 > 0$ such that, for $\epsilon < \epsilon_0$, Φ has an SRB measure μ. Furthermore, μ is invariant and exponentially mixing under space-time translations: there exist $\theta > 0$, $C < \infty$, such that, for any two finite subsets $B, D \subset \mathbb{Z}^d$ of the lattice and for any functions $g \in L^\infty(\mathcal{M}_B)$, $h \in C^\alpha(\mathcal{M}_D)$*

$$\left| \int g \circ f^{(n)} h \, d\mu - \int g \, d\mu \int h \, d\mu \right| < C \|g\|_\infty \|h\|_\alpha e^{-\theta(n+d(B,D))}, \tag{2.4}$$

where $d(B, D)$ is the distance between the subsets B and D, and C depends on the diameters $d(B)$, $d(D)$ of B and D.

The property (2.4) was introduced in Bunimovich & Sinai (1988) and was called there "space-time chaos". (In fact, the space-time chaos in that paper was defined as a decay of space-time correlations, rather than as the exponential decay of these correlations as in (2.4). As we already mentioned, it is better to call this property "space-time mixing" and to call (2.4) "space-time exponential mixing".)

In Bunimovich & Sinai (1988) this property was proven for one dimensional lattices ($d = 1$) of one-dimensional expanding maps of the unit interval with some special type of weak spatial interactions. Bricmont & Kupiainen (1996) deal with a more general class of spatial interactions than the one considered in Theorem 2.1. The existence of SBR measures for weak interactions of general type on $\mathbb{Z}^1$ was proven in Keller & Künzle (1992). The results on the existence of SBR measures for CML on lattices of arbitrary (finite) dimension were obtained also in Volevich (1993 & 1994).

SBR measures for weakly interacting hyperbolic diffeomorphisms on $\mathbb{Z}^1$ were studied in Pesin & Sinai (1991). They assumed that the local map f had a hyperbolic attractor Λ (i.e., there exists an open neighborhood $U \supset \Lambda$ such that $\overline{f(U)} \subset U$ and $\Lambda = \bigcap_{n \geq 0} f^n(U)$, where $\bar{B}$ denotes the closure of a set B). In Jiang (1995) the more general situation was studied when a diffeomorphism f has a locally maximal hyperbolic set Λ (i.e., there exists an open set $O \supset \Lambda$ such that $\Lambda = \bigcap_{n \in \mathbb{Z}} f^n(\bar{O})$).

The problem of uniqueness for the corresponding SBR measures was addressed in Bricmont & Kupiainen (1995, 1996, 1997), Jiang (1995), and

Jiang & Mazel (1996). These papers have employed and adapted for LDS the method of cluster or polymer expansions developed for the lattice systems of statistical mechanics (Malyshev & Minlos, 1991; Simon, 1994). We address the interested reader to Bricmont & Kupiainen (1997) and Jiang & Mazel (1996), which contain a nice and clear description of these problems. (Some similar results were claimed in the series of papers by Gundlach & Rand (1993a, b, & c). However, most of the main results of these papers are wrong. The corresponding explanations and counterexamples can be found in Jiang (1995).)

The above mentioned results were obtained by the application and development of an appropriate thermodynamic formalism, which serves as the most powerful tool to study statistical properties of (nonextended) dynamical systems (Sinai, 1972 & 1996; Bowen, 1975; Ruelle, 1978). It has been pointed out by Bunimovich & Sinai (1988 & 1993) that the results on the existence and uniqueness of SBR invariant measures for LDS with weak interactions correspond to the results on the absence of phase transition for high temperatures in lattice spin models of statistical mechanics. In fact, the region of weak spatial interactions in LDS corresponds via the thermodynamic formalism to the high-temperature region in the spin models.

It was pointed out by Bunimovich & Sinai (1988) that lattice models of statistical mechanics that correspond to LDS are $(d+1)$-dimensional, where $d \geq 1$ is the dimension of the lattice. On the other hand, the lattice spin models that correspond to the (nonextended) finite dimensional hyperbolic dynamical systems are the one-dimensional ones and do not demonstrate phase transitions. Different phases, however, are possible in lattice models with two or more dimensions (Dobrushin, 1968).

Therefore, it has been suggested by Bunimovich & Sinai (1988) that such phase transitions can be considered as the appearance of coherent structures in the corresponding extended dynamical systems (LDS).

This idea has received much attention. However, so far the results are only numerical and far from complete (see e.g., Bunimovich *et al.*, 1990; Miller & Huse, 1993; Boldrighini *et al.*, 1995; Egolf & Greenside, 1995). The idea of Bunimovich & Sinai (1988) on the connection of phase transitions with the appearance of coherent structures in LDS seems to be confirmed in these numerical studies, but the problem of the character of such transitions is completely open.

We will briefly discuss a recent result on the existence of such phase transitions (Bunimovich & Carlen, 1995). Consider the following CML:

$$x_i(n+1) = f(x_i(n)) + \frac{\epsilon\, \alpha(x_i(n))}{2}\left(f(x_{i-1}(n)) - 2f(x_i(n)) + f(x_{i+1}(n))\right), \tag{2.5}$$

where $x_i \in [0,1]$, $f(x) = 4x(1-x)$, $i \in \mathbb{Z}$ and $\alpha : [0,1] \to [0,1]$ is a function of class C^2, $\alpha(0) = \alpha(1) = B$ and $\alpha(x) > 0$ for $x \in]0,1[$. For brevity, we will also write (2.5) as

$$\mathbf{x}(n+1) = \Phi\mathbf{x}(n).$$

This CML was studied in Volevich (1991) where, for sufficiently small ϵ, the existence of a space-time mixing SBR measure μ was proven. Moreover, it was shown there that some class of nonequilibrium (noninvariant) probabilistic measures ν converge (weakly) to μ under the action of dynamics Φ, i.e., $(\Phi^n)^*\nu \to \mu$ as $n \to \infty$.

It is proven by Bunimovich & Carlen (1995) that when the amplitude ϵ of spatial interactions becomes sufficiently large some measures ν of this class evolve under the action of $\{\Phi^n\}$ to measures that differ from μ.

In this sense the LDS (2.5) demonstrates the existence of phase transitions. The corresponding coherent structures are travelling waves. However, it is likely that between the range of ϵ that corresponds to space-time chaos and the range of the large spatial interactions studied by Bunimovich & Carlen (1995), some other phase transitions occur. Therefore this result sheds no light on the character of the corresponding transitions. So this area, while being very attractive, is currently completely open.

3. An Onset of Chaos in CML via the Peak-Crossing Bifurcation

In this section we discuss the mechanism that causes the onset of chaos in some classes of CML with a simple local dynamics. All these systems are of the reaction-diffusion type:

$$x_{\mathbf{i}}(n+1) = (\Phi x(n))_{\mathbf{i}} = f(x_{\mathbf{i}}(n)) + \frac{\epsilon}{2d} \sum_{\|\mathbf{j}-\mathbf{i}\|=1} (f(x_{\mathbf{j}}(n)) - f(x_{\mathbf{i}}(n))) \quad (3.1)$$

where $\mathbf{i} \in \mathbb{Z}^d$, $n \in \mathbb{Z}_+$, $\epsilon > 0$.

The basic (simplest) class of local maps is formed by piecewise linear one-dimensional maps $f : [0,1] \to [0,1]$ of a special type. Such maps are depicted in Fig. 1. In what follows we will refer mainly to the continuous map (Fig. 1b).

The characteristic property of this map is the existence of some flat components where f assumes the same value. The slopes of the non-flat linear components of a graph of f are given by λ, ω, α, β in that order. We denote by F the region of $[0,1]$ over which the flat part is defined, namely, $F = [P,A] \cup [C,1]$. The height of the flat part is given by the parameter α. We shall call the segment $[A,C]$ the "tent region." Denote by B the point where f takes the maximum value (apex) in the tent region and set $b := f(B)$.

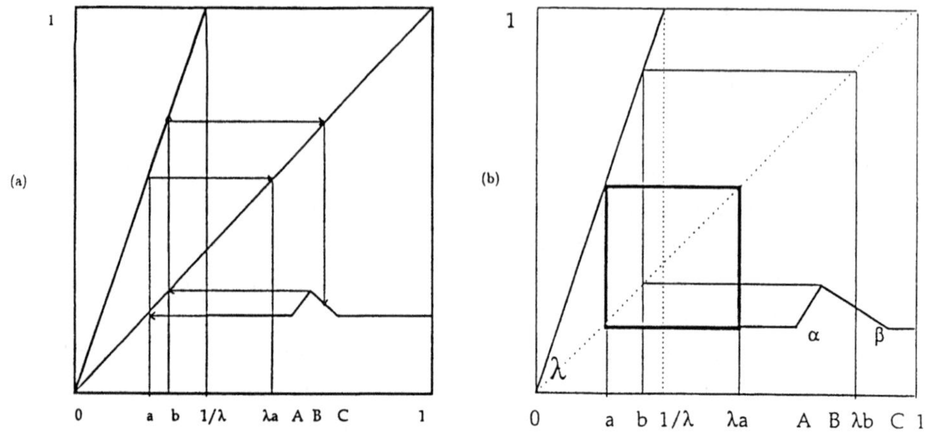

Figure 1. The basic map f

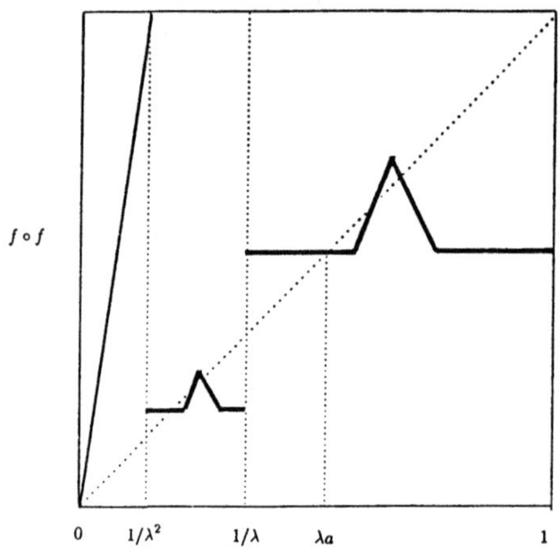

Figure 2. The second iterate of f.

This local map f appears as a modification of the trivial maps which served as the local maps for the CMLs discussed in Bunimovich *et al.* (1992). The characteristic feature of our map f (as well as of the one discussed in Bunimovich *et al.*, 1992) is the existence of a flat region that carries the superstable periodic trajectory $\{a, f(a)\}$. This trajectory attracts almost all points of the segment $[0, 1]$, and this causes the very simple (trivial) dynamics of the local map.

The crucial difference is that our local map also has the "tent region" $[A, C]$ situated inside its flat part. It is easy to see from Fig. 1 that such a map depends on seven parameters (e.g., on λ, a, b, P, A, B, C). However, we believe that it is the simplest example of CML that demonstrates the onset of chaotic motion after the spatial interactions exceed some (non-zero) threshold.

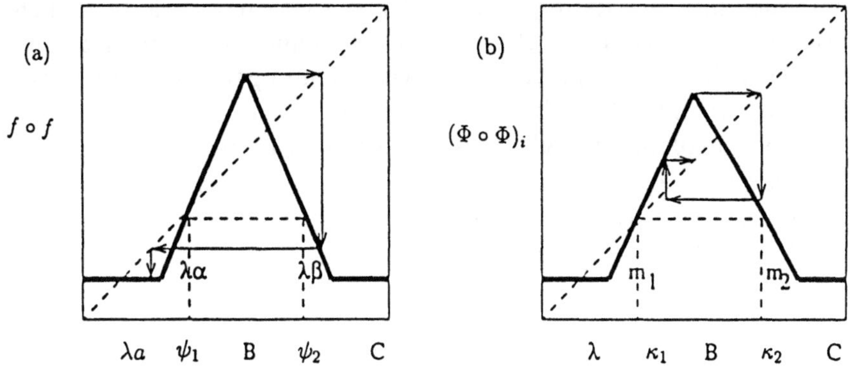

Figure 3. A peak-crossing bifurcation. $\lambda\alpha$, $\lambda\beta$, m_1, and m_2 denote slopes.

It is easy to see that a necessary condition for the coupled map Φ to have nontrivial dynamics is that the image of the tent region under some iterate of f intersects the diagonal of the unit square (Fig. 2). Otherwise, almost every trajectory gets "glued" into the superstable periodic orbit. For the sake of brevity we assume that the second iterate $f \circ f$ satisfies this condition (Fig. 2).

The idea of our mechanism is suggested by the bifurcation that is familiar to everybody who has dealt with families of one-dimensional maps. If some peak on a graph of the function $f : [0,1] \to [0,1]$ intersects the diagonal then two situations can occur (Fig. 3). The first one (depicted in Fig. 3a) refers to the case when the image of the tip leaves the peak region under the iterations of f. In the second case the images of the tip return to the peak region, cross it and stay there forever (Fig. 3b); then in the peak region there is some invariant set A with positive Lebesgue measure, such that the dynamics $f|A$ is chaotic.

Suppose that $f \circ f$ has the graph as in Fig. 3a, corresponding to simple dynamics. The idea is that under the action of the perturbations produced by the spatial interactions of the local map at a site i with its neighbors, the map induced at this site by the CML $(\Phi \circ \Phi)$ will become like the one in Fig. 3b.

It is worthwhile to mention that this situation is a new one from the point of the general theory of bifurcations. Indeed, we consider local maps that all are identical (all their internal parameters coincide), and in addition the system is translationally invariant. However, spatial interactions (in some range of their strength) induce the occurrence of peak-crossing bifurcations at some sites of the lattice $\mathbb{Z}^d$. As the result the lattice gets partitioned into alternating clusters of sites that carry periodic and chaotic motion respectively. Thus, the emerging dynamics is characterized by spatial intermittency. The partition into chaotic and periodic clusters depends on the initial state of the lattice (the initial conditions).

We will explain now why this scenario can be realized. The complete proofs can be found in Bunimovich & Venkatagiri (1996).

Suppose that $x_i(0)$ belongs to the tent region, while $x_{i+j}(0) \in F$, where $j = \pm 1, \pm 2, \pm 3, \pm 4$. One can write

$$\begin{aligned}
x_i(1) &= (1-\epsilon)f(x_i(0)) + \tfrac{\epsilon}{2}\left(f(x_{i-1}(0)) + f(x_{i+1}(0))\right) \\
&= (1-\epsilon)f(x_i(0)) + \epsilon a \\
x_{i-1}(1) = x_{i+1}(1) &= (1-\epsilon)a + \tfrac{\epsilon}{2}\left(a + f(x_i(0))\right) \\
&= \left(1-\tfrac{\epsilon}{2}\right)a + \tfrac{\epsilon}{2} f(x_i(0)) \\
x_i(2) = (\Phi \mathbf{x}(1))_i &= (\Phi \circ \Phi \mathbf{x}(0))_i \\
&= (1-\epsilon)f(x_i(1)) + \tfrac{\epsilon}{2}\left(f(x_{i-1}(1)) + f(x_{i+1}(1))\right)
\end{aligned}$$

Thus all of the terms in the expression for $(\Phi \circ \Phi \mathbf{x}(0))_i$ depend on $x_i(0)$. Thus, if the two nearest neighbors of x_i keep returning to F after every two iterations, then the map $(\phi \circ \phi x)_i$ can be defined solely in terms of x_i. Therefore this map is one-dimensional, and we can look for conditions that ensure it to be as in Fig. 3b while $f \circ f$ is as in Fig. 3a.

We denote by $(\Phi \circ \Phi)_i$ the action of the second iterate of Φ restricted to the site i. It is important to mention that the dynamical system generated by the map $(\Phi \circ \Phi)_i$ is an autonomous one (i.e., its evolution depends only upon the initial condition at ith site, $x_i(0)$) because of the existence of the flat region for the map f. It causes a complete "loss of the memory on perturbation" by the sites that follow the superstable periodic trajectory (i.e., values of $x_j(n)$ at these sites keep returning to F at each other moment of time).

Let ψ_1 be the minimal fixed point of $f \circ f$ in the tent region (see Fig. 3a). We require the following:

$$f \circ f(B) > B \tag{3.2}$$

$$f^{(4)}(B) < \psi_1 \tag{3.3}$$

Also we suppose that the slopes m_1 and m_2 are greater than 1. Denote by κ the minimal fixed point of $(\Phi \circ \Phi)_i$ in the tent region (Fig. 3b).

In order to ensure that $(\Phi \circ \Phi)_i$ has the shape as in this fig. we also require the following:

$$(\Phi \circ \Phi)_i(B) > B, \text{ and} \tag{3.4}$$

$$(\Phi \circ \Phi)_i((\Phi \circ \Phi)_i(B)) = \Phi_i^{(4)}(B) > \kappa. \tag{3.5}$$

Let κ_1 be the point in the interval $[B, C]$ such that $(\Phi \circ \Phi)_i(\kappa_1) = \kappa$. We will call the interval (κ, κ_1) the peak region. As depicted in Fig. 3, we demand that under the action of the perturbations produced by coupling with the neighboring sites, the dynamics at the site i changes. This change is brought about by the image under $(\Phi \circ \Phi)_i$ of the point B (which gives the maximum value of f in the tent region) crossing over to the peak region under sufficient coupling. Further iterates under $(\Phi \circ \Phi)_i$ are trapped within the peak region.

We assume that $\lambda > 1$ and $\frac{1}{\lambda^2} < a < \frac{1}{\lambda}$. The last condition ensures that the superstable trajectory has period 2. All other periods can be treated analogously. To define a map f we now specify the ranges for the other parameters $a < b < 1/\lambda$, $1/\lambda < p < \lambda a$, $\lambda a < A < \lambda b$, $A < B < \lambda b$, $\lambda b < C < 1$.

We define the (positive) slopes of the piecewise linear parts of f as follows:

$$\omega := (1-a)/(P - 1/\lambda), \quad \alpha := (b-a)/(B-A), \quad \beta := (b-a)/(C-B).$$

Finally the local map is defined as follows:

$$f(x) = \begin{cases} \lambda x & 0 \le x \le 1/\lambda \\ 1 - (x - 1/\lambda)\omega & 1/\lambda < x < P \\ a & P \le x \le A \\ a + (x - A)\alpha & A < x \le B \\ a - (x - C)\beta & B < x < C \\ a & C \le x \le 1 \end{cases}$$

The following result is proven in Bunimovich & Venkatagiri (1996).

Theorem 3.1. *There exists an open region in the parameter space of the local map f and an interval of values of the strength of spatial interactions $(\epsilon_1^{(d)}, \epsilon_2^{(d)}), \epsilon_1^{(d)} > 0$, such that for a generic (with respect to the product measure μ) initial state $\mathbf{x}(0) \in \mathcal{M}$ the lattice $\mathbb{Z}^d$ contains infinitely many sites with chaotic motion surrounded by sites with F-recurrent motion.*

Here by a chaotic motion we mean that the map induced by $(\Phi \circ \Phi)_\mathbf{i}$ admits an absolutely continuous ergodic invariant measure at infinitely many sites $\mathbf{i} \in \mathbb{Z}^d$. Moreover, this map has a continuous countable (more precisely, the Lebesgue) component in the spectrum. The remaining sites return to the flat region F after every two iterations.

The next natural step is to prove the same statement for the analogous class of LDS rather that only for CML (in progress). We believe that this mechanism of the onset of chaos via the peak crossing bifurcations is rather general. It should work also for local smooth maps that have an "almost" flat region in a neighborhood of some local maximum (or minimum). However, for polynomial maps f it might work only if the degree of f is big enough. This makes the corresponding analysis complicated; even for the piecewise linear map the algebra was rather cumbersome.

4. On Stability of Structures in Extended Systems

To study the chaos-order transitions in LDS it is necessary to develop the theory of stability of their solutions. The problem is that LDS are infinite dimensional dynamical systems. Therefore one should choose a proper norm in a corresponding functional space. This problem has been addressed in many papers (see e.g., Afraimovich & Bunimovich, 1993; Afraimovich & Chow, 1995; Afraimovich *et al.*, 1994; Afraimovich & Nekorkin, 1994; Chow & Shen, 1995; Shen, 1995). We will discuss in this section a general result on nonlinear stability of simple solutions in LDS (Bunimovich & Carlen, 1995).

Let

$$B^0 = \{u = \{u_{\mathbf{j}}\}, \mathbf{j} \in \mathbb{Z}^d, u_{\mathbf{j}} \in \mathbb{R}^m \text{ vanishing for all but finitely many } j\},$$

and where $\mathbb{R}^m$ is m-dimensional Euclidean space with the standard inner product and the norm.

We shall use the following two norms on B^0:

(1) The ℓ^2-norm, $\|\cdot\|_{\ell^2}$, that is generated by the inner product $\langle \mathbf{u}, \mathbf{v}\rangle = \sum_{\mathbf{j}\in\mathbb{Z}^d}(u_{\mathbf{j}}, v_{\mathbf{j}})$.

(2) For each $q > 0$, we define the q-norm $\|\cdot\|_q$, to be the norm generated by the inner product

$$\langle \mathbf{u}, \mathbf{v}\rangle_q = \sum_{\mathbf{j}\in\mathbb{Z}^d} \frac{(u_{\mathbf{j}}, v_{\mathbf{j}})}{q^{|\mathbf{j}|}},$$

where $|\mathbf{j}| = \max\{|j_1|, \ldots, |j_d|\}$.

Certainly, the ℓ^2 norm is just the special case $q = 1$ of the q-norm that we have singled out for the obvious reasons. The q-norms for $q > 1$ have been used by Afraimovich *et al.* (1994) and Afraimovich & Nekorkin (1994) to study the stability of travelling waves in some CML, and the q-norms for $q < 1$ have been used by Bunimovich & Sinai (1988) to study hyperbolic properties of some other CML. The paper by Bunimovich & Carlen (1995) has unified these approaches.

Also, we shall consider the space B of all uniformly bounded sequences equipped with the ℓ^∞ norm, $\|\mathbf{u}\|_{\ell^\infty} = \sup_{\mathbf{j}\in\mathbb{Z}^d}|u_{\mathbf{j}}|$.

It is easy to see that

$$\|\mathbf{u}\|_{q_1} \geq \|\mathbf{u}\|_{q_2} \quad \text{for all } \mathbf{u} \in B^0, \tag{4.1}$$

whenever $q_1 \leq q_2$.

Another useful inequality (Bunimovich & Carlen, 1988) is

$$|u_{\mathbf{j}}| \leq \|\mathbf{u}\| q^{|\mathbf{j}|/2}. \tag{4.2}$$

In particular,

$$\|\mathbf{u}\|_{\ell^\infty} \leq \|\mathbf{u}\|_q \quad \text{for all } 0 < q < 1 \tag{4.3}$$

We denote by B_q the completion of B^0 under the q-norm.

The temporal evolution of our LDS is to be generated by the map $\Phi : B_q \to (\mathbb{R}^m)^{\mathbb{Z}^d}$ defined by the relation

$$(\Phi\mathbf{u})_{\mathbf{j}} = u_{\mathbf{j}} + \mathcal{F} \circ S_{\mathbf{j}}(\mathbf{u}), \tag{4.4}$$

where $\mathcal{F} : B_q \to \mathbb{R}^m$ is a differentiable map of class C^2, bounded in a suitable sense, and the spatial translations $\{S_{\mathbf{j}}\}_{\mathbf{j}\in\mathbb{Z}^d}$ are defined via

$$(S_{\mathbf{j_o}}\mathbf{u})_j = u_{\mathbf{j}-\mathbf{j_o}} \tag{4.5}$$

The operators $S_{\mathbf{j}}$ are bounded, linear and commute with the evolution operator Φ (Afraimovich & Nekorkin, 1994; Afraimovich *et al.*, 1994; Afraimovich & Chow, 1995).

We suppose that for some integer r, called the range of the interactions, $\mathcal{F}$ does not depend on the coordinates $u_{\mathbf{i}}$ for $|\mathbf{i}| < r$. Thus if we define $\{u_{\mathbf{j}}\}^r \in (\mathbb{R}^m)^{(2r+1)^d}$ by $\{u_{\mathbf{j}}\}^r = \{u_{\mathbf{i}} : \ |\mathbf{i}-\mathbf{j}| \leq r\}$ then we can write

$$(\Phi\mathbf{u})_{\mathbf{j}} = u_{\mathbf{j}} + G(\{u_{\mathbf{j}}\}^r) \tag{4.6}$$

where $G : (\mathbb{R}^m)^{(2r+1)^d} \to \mathbb{R}^m$ is a differentiable map of class C^2_B, i.e., uniformly bounded with uniformly bounded derivatives of order two.

It can be shown (Afraimovich & Nekorkin, 1994; Afraimovich *et al.*, 1994) that under some mild conditions that depend on the q-norm under consideration that $\Phi(B_q) \subset B_q$ for $q > 1$. Thus the dynamical system $\{\Phi^r\}$ generated by Φ is defined on B_q and one can write

$$\mathbf{u}(n+1) = \Phi(\mathbf{u}(n)) \tag{4.7}$$

Now suppose we have a space time coherent solution $\mathbf{u}$ to (4.4); this means that for some $k > 0$, $\ell > 0$ and

$$\Phi^k\mathbf{u} = \mathbf{u} \tag{4.8}$$

$$S_\ell\mathbf{u} = \mathbf{u}_k \tag{4.9}$$

Substituting Φ^k for Φ, we may reduce to the case $k = 1$. Notice that the range of Φ^k is kr if the range of Φ is r.

We now rewrite (4.8) more explicitly as

$$u_j^{n+1} = (\Phi u^n)_j = u_j^n + G(u_{j-r}^n, \ldots, u_{j+r}^n) \tag{4.10}$$

Hence, if we introduce $\xi_j^n = u_j^n - u^j$, with $\mathbf{u}$ solving (4.8), we have

$$\xi_j^{n+1} = (A\xi^n)_j + B(\xi^n)_j, \tag{4.11}$$

where

$$(A\xi^n)_j = \sum_{\ell=-r}^{r} \left(\delta_{\ell,0} + \frac{\partial G}{\partial x_\ell}(\{\mathbf{u}\}^r) \right) \xi_{j+\ell}^n \tag{4.12}$$

is the formal linearization of (4.10).

The difficulty with this is that the operator A is not the Fréchet derivative of Φ in the q-norm, although it is the ℓ^2-norm. This is because the nonlinear correction $B(\xi)_j$ in (4.11) is small compared to $(A\xi)_j$ only when the sup-norm of ξ is small. But the q-norm does not control the sup norm. A useful fact, however, is that A is strictly contractive in q-norm (Afraimovich & Bunimovich, 1993). Moreover, under some readily checked conditions A is a strict contraction for all q in some open interval about $q = 1$ (Bunimovich & Carlen, 1995).

Let $T(q)$ denote A^*A where the adjoint operator A^* of A is taken with respect to the B_q inner product. Define

$$\alpha := \max\{[T(q)]_{i,i} : \ 0 \leq i \leq \ell\} \tag{4.13}$$

and

$$\beta := \max \left\{ \sum_{j \neq i} | \ [T(q)]_{i,j} \ |: \ 0 \leq i \leq \ell \right\} \tag{4.14}$$

where ℓ is the spatial period of the solution $\mathbf{u}$.

We introduce the Hessian of G, Hess(G), a map from $\mathbb{R}^{(2r+1)}$ to the symmetric $(2r+1) \times (2r+1)$ matrices, by $[\text{Hess}(G)(x)]_{k,\ell} = \frac{\partial^2}{\partial x_k \partial x_\ell} G(x)$. Denote by $\|\text{Hess}(G)(x)\|_{\text{H.S.}}$ the Hilbert-Schmidt norm of Hess$(G)(x)$:

$$\|\text{Hess}(G)(x)\|_{\text{H.S.}}^2 = \sum_{|k|,|\ell| \leq r} \left(\frac{\partial^2}{\partial x_k \partial x_\ell} G(x) \right)^2 \tag{4.15}$$

Theorem 4.1. (Bunimovich & Carlen, 1995) *Suppose that the linearized evolution operator A is such that $\alpha + \beta \leq 1$. Then for all q with*

$$1 \leq q \leq q_0 \left(\frac{\alpha + \beta + 1}{2(\alpha + \beta)} \right)^{1/r}$$

we have that

$$\|\xi^n\|_q \leq \|\xi^n\|_{1/q} \leq \left(\frac{\alpha+\beta+3}{4}\right)^{n/2} \|\xi^0\|_{1/q}$$

whenever

$$(2r+1)^2 q^r (\sup\{\|Hess(G)(x)\|_{\text{H.S.}} \mid x \in \mathbb{R}^{m(2r+1)}\})\|\xi^0\|_{1/q}^2 \leq \frac{1-(\alpha+\beta)}{2}$$

This theorem provides readily checked stability conditions for coherent solutions of LDS. A similar analysis leads to ℓ^∞ stability (Bunimovich & Carlen, 1995). Analogously, the conditions for stability of the solutions of lattices of coupled ODEs can be obtained, when these systems are of the form

$$\frac{du_{\mathbf{j}}}{dt} = f(\{u_{\mathbf{j}}\}^r), \tag{4.16}$$

with $f : (\mathbb{R}^m)^{(2r+1)^d} \to \mathbb{R}^m$ a differentiable map of class C^1 such that $|\partial f/\partial u_{\mathbf{i}}| \leq M < \infty$ uniformly and for all $\mathbf{i}$, and such that

$$\sup_{\mathbf{j}\in\mathbb{Z}^d} \{\|f\{u_{\mathbf{j}}\}^r\|\} \leq N < \infty.$$

See Bunimovich & Carlen (1995) for details.

An important problem concerns the study of mechanisms that cause the appearance (or the disappearance) of spatial structures in extended systems.

Any (spatial) structure will also contain some structures with smaller scales. We will call these smaller structures (spatial) patterns. The simplest spatial structures contain only one spatial pattern. For example, one encounters such a situation in sinusoidal and other simple periodic structures. However, generally a spatial structure contains patterns with several different scales.

It has been proposed in Bunimovich (1995) that the stability of a spatial structure is determined, at least in some cases, by the stability of its most tiny pattern, i.e., of the pattern with the minimal spatial scale. We will discuss briefly the results of some numerical simulations (Bunimovich *et al.*, 1997) that confirm this claim.

Consider the CML generated by the nearest-neighbor diffusive interactions of (local) logistic maps. The phase space of the dynamical system under study is, geometrically, the N-dimensional cube. Thus a point $\mathbf{x} \in \mathcal{M}$

is an N-dimensional vector $\mathbf{x} = (x_1, x_2, \dots, x_N)$ where $0 \le x_i \le 1$. The dynamics on $\mathcal{M}$ is generated by the mapping $\Phi_{N,a,\epsilon} : \mathcal{M} \to \mathcal{M}$ defined as

$$x_i^{(n+1)} = \left(\Phi_{N,a,\epsilon}(\mathbf{x}^{(n)})\right)_i = f(x_i^{(n)}) + \frac{\epsilon}{2}\left(f(x_{i-1}^{(n)}) - 2f(x_i^{(n)}) + f(x_{i+1}^{(n)})\right) \tag{4.17}$$

where $1 \le i \le N$ is the space coordinate, n is the discrete time, and $f(u) = au(1-u)$, $0 < a \le 4$, is the logistic map. Periodic boundary conditions are always assumed, i.e., $x_i^{(n)} \equiv x_{i+N}^{(n)}$ for any i and N. The simulations were made for the lattices with the sizes up to $N = 32$.

To verify the principle formulated above one should study the simplest nontrivial spatial structures. The natural choice is formed by kinks because only homogeneous structures are simpler than kinks.

One easily gets kinks for the coupled logistic maps. Let us put $\epsilon = 0$. It is well known that the logistic map $f(u)$ has an unstable trajectory of period 2 for $a > a_1 = 3.4494$. Denote the points of this trajectory by z_0 and z_1, $0 < z_0 < z_1 < 1$. Obviously one gets 2^N solutions $(x_1, \dots, x_N)$ for $\Phi_{N,R,0}$ by taking $x_i = \{z_0, z_1\}$, $i = 1, 2, \dots, N$, and all of them have period two in time. These solutions are all unstable and all but two contain kinks (Giberti & Vernia, 1994).

We are interested in whether or not these periodic solutions acquire stability when one switches on the spatial interactions ($\epsilon > 0$) and in their loss of stability when the strength of spatial interactions (ϵ) or the nonlinearity (a) become too strong.

It is easy to show that all 2^N two-periodic cycles of the map $\Phi_{N,a,0}$ are unstable when $a > a_1$. For the sake of simplicity we will denote a cycle of $\Phi_{N,a,0}$ of this type by a sequence of 0's and 1's, where 0 stands for z_0 and 1 stands for z_1. Obviously, for $\epsilon > 0$, a sequence of z_0's and z_1's doesn't form a solution of (4.17) any more, but we will still keep the same coding. This coding makes sense until the corresponding solution, continuously prolonged from $\epsilon = 0$, gets destroyed for some $\epsilon > 0$.

Our coding of the two-periodic cycles of $\Phi_{N,a,0}$ is not unique. Indeed, one gets the same cycle by changing all 0's into 1's and vice versa, and a symmetric, and thus equivalent cycle, by rotation. In order to have a unique representation we take the one that corresponds to the minimal integer.

We define fundamental cycles as those that contain exactly one kink. Thus, the fundamental cycles are the ones that have a code consisting of exactly one sequence of 0's and one sequence of 1's. Hence the list of fundamental cycles contains 01, 001, 0001, 0011, 00001, 00011 and so on.

Denote by $\mathcal{F}_{n,m}$ the fundamental cycle of length n with m 1's. Obviously, $\mathcal{F}_{n,m} = \mathcal{F}_{n,n-m}$. Therefore we can assume that $m \le n/2$.

We are interested in the spatial structures (solutions of (4.17)) that contain many kinks. The fundamental cycles form the elements of such

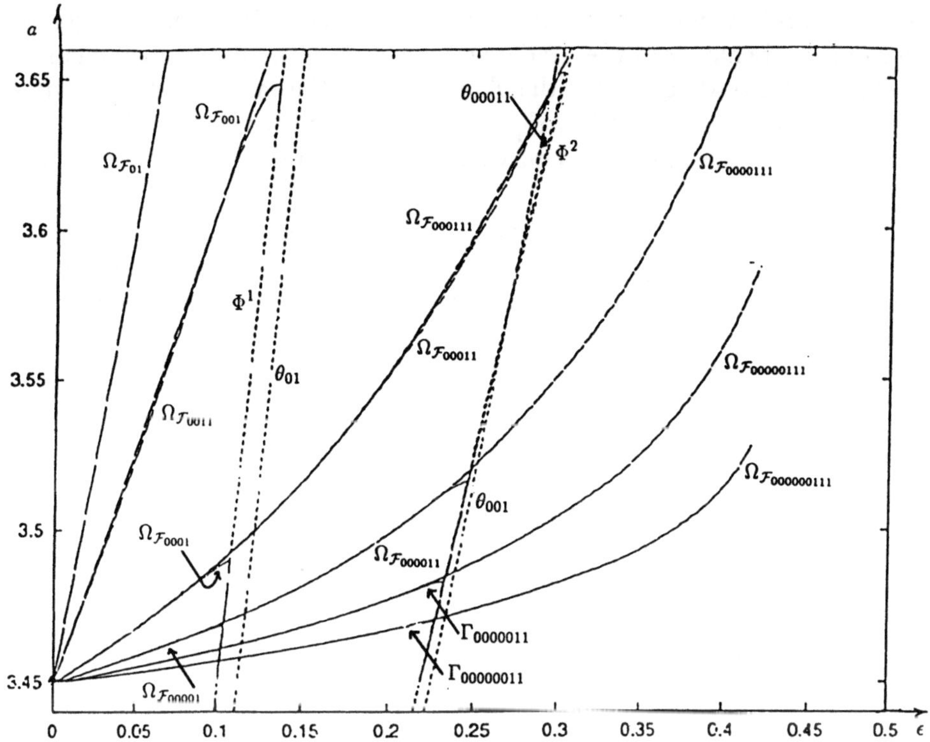

Figure 4. Regions of stability of the shortest fundamental cycles. The bifurcation curves are broken (continuous) if a cycle loses (acquires) stability

structures. However, these elements are not the simplest ones because each fundamental cycle consists of two homogeneous clusters. Such homogeneous clusters will be called patterns. It is easy to see that any nonfundamental cycle C_n (a spatial structure of length n) can be decomposed into fundamental cycles, e.g., $0001011000111001 \equiv \mathcal{F}_{4,1} \oplus \mathcal{F}_{3,2} \oplus \mathcal{F}_{6,3} \oplus \mathcal{F}_{3,1}$.

One of the main observations made in Bunimovich *et al.* (1997) is that for a fixed $a > a_1$ any structure C_n acquires stability for some $\epsilon_1 = \epsilon(C_n)$ if all the fundamental cycles $\mathcal{F}_{n_i,m_i} \subset C_n$ acquire stability and the intersection of their stability ranges is not empty. Approximately, this intersection represents the stability range of the entire structure. (The rigorous proof of this fact is given by Bunimovich & Carlen (1996) for some more general classes of spatial structures.)

Therefore, this result reveals that the stability of a spatial structure that is decomposable into simpler structures, is determined by the stability of these smaller components.

We will show now that, actually, the stability of a structure is defined by the stability of its pattern with the smallest length. In order to do so, we consider the behavior of the fundamental cycles as the parameters change.

The stability diagram for the shortest (and the most important) fundamental cycles is depicted in Fig. 4. In this figure the dashed line $\Omega_{\mathcal{F}_{n,m}}$ corresponds to the bifurcation curve on which the fundamental cycle $\mathcal{F}_{n,m}$ acquires stability.

It has been found that any cycle including 01 or/and 001 disappears inside a narrow strip on the parameters plane approximately bounded by the curves θ_{01} and θ_{001}. Only cycles not including 01 and 001 can survive for larger ϵ.

The next set of composed cycles are the ones for which all the component patterns have at least length two, with at least one pattern exactly of length two. All these cycles disappear in a strip which includes θ_{0011}.

These results (as well as many others obtained in Bunimovich *et al.*, 1997) confirm the principle of the destruction of a spatial structure through its pattern of the smallest scale. The natural problem is to find the limits of applicability of this principle.

5. Local Nonlinearity and Spatial Complexity in LDS

We have mentioned and used several times the fact that a (semi)group generated by space-time shifts (Φ, S) acts on the phase space $\mathcal{M}$ of the LDS. However, it is in fact only the time dynamics $\{\Phi^n\}$ that really governs the behavior of the extended system. The group of space shifts $\{S_{\mathbf{j}}\}$, $\mathbf{j} \in \mathbb{Z}^d$, rather acts in the heads of the investigators than on the real systems. Therefore, it is natural to study how spatial structures are actually organized in extended dynamical systems under the action of time dynamics (Coullet & Huerre, 1990).

We will discuss in this section the relation between local production of entropy and the complexity of spatial structures in some LDS of the reaction-diffusion type.

Consider a discrete version of the nonlinear diffusion equation

$$u_j(n+1) = u_j(n) + \alpha f(u_j(n)) + \beta(u_{j-1}(n) - 2u_j(n) + u_{j+1}(n)) \quad (5.1)$$

where $j \in \mathbb{Z}^1$, $n \in \mathbb{Z}_+$, $\alpha, \beta > 0$ and $f : \mathbb{R} \to \mathbb{R}$ is a local map.

We will be interested in static solutions (fixed points) of the LDS (5.1). The equation that defines fixed points of (5.1) has the form

$$u_{j+1} = -\frac{\alpha}{\beta} f(u_j) + 2u_j - u_{j-1} \quad (5.2)$$

Denoting $x_j = u_{j-1}$, $y_j = u_j$ one can rewrite (5.2) in the form

$$\begin{aligned} x_{j+1} &= y_j \\ y_{j+1} &= -\tfrac{\alpha}{\beta} f(y_j) + 2y_j - x_j \end{aligned} \quad (5.3)$$

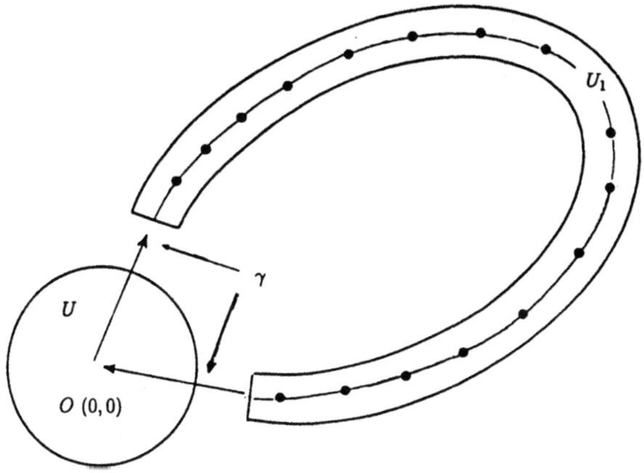

Figure 5. A homoclinic trajectory and its neighbourhood

The dynamical system (5.3) is defined on the set of static solutions of (5.1) and is generated by the spatial shifts. Indeed, in (5.3) the index j plays the role of a time coordinate. Suppose that $f(0) = 0$ and the origin $x = y = 0$ is a fixed point of the system (5.1). The characteristic equation at 0 has the form $\lambda^2 - (2 + \lambda_0)\lambda + 1 = 0$ where $\lambda_0 = -\frac{\alpha}{\beta} f'(0)$. Thus

$$\lambda_{1,2} = 1 + \frac{\lambda_0}{2} \pm \sqrt{(1 + \lambda_0/2)^2 - 1} \tag{5.4}$$

If $\lambda_0 > 0$ then the point $0 = (0,0)$ is of saddle type, i.e., $0 < \lambda_2 < 1 < \lambda_1$. We assume that there exists a homoclinic trajectory for the point 0 that is stable with respect to the system (5.1). (The corresponding sufficient conditions can be found in Afraimovich & Bunimovich (1995).)

Fix now some ϵ-neighborhood U of the point 0. If ϵ is sufficiently small then only a finite number r of points of the homoclinic trajectory are situated outside U. Consider some neighbourhood U_1 of this finite set of points of the homoclinic trajectory that does not intersect with U (Fig. 5).

Recall now that the system (5.3) describes the "dynamics" in space, i.e., in the lattice $\mathbb{Z}^1$. Therefore, we can say that a solution of (5.3) has a "hump" (burst) if it intersects U_1. The distribution of such bursts characterizes a spatial complexity of the corresponding solution, that, in fact, is the "frozen" spatial structure of the LDS (5.1).

The homoclinic solution spends r "steps of dynamics" (that are actually steps in space $\mathbb{Z}^1$) outside U making these steps along the finite set of r points that are situated in U_1. After this passage (that represents a "hump" in space) it gets back to U and never leaves this neighborhood again.

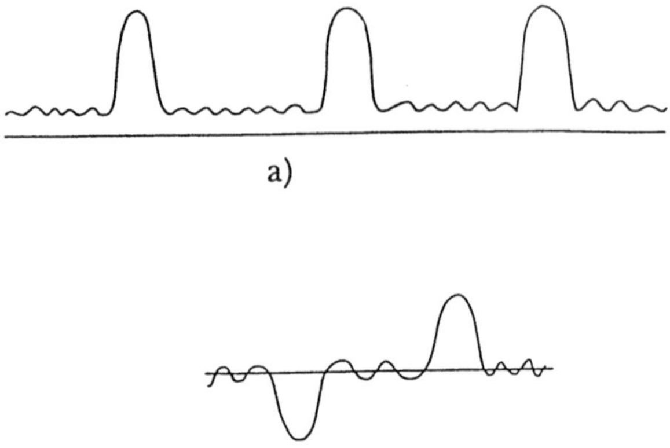

a)

b)

Figure 6. Spatial structures with "humps"

There exists however an infinite set of static solutions of (5.1) that have any finite number of such passages through U_1 and visit U between these passages. The corresponding solutions have a form depicted in Fig. 6. One can think of U as the "background" of the spatial field (on a lattice) if ϵ is small enough.

In other words, one can find an ϵ_1-neighborhood U_1 such that an infinite collection of trajectories belonging to $U \cup U_1$ can be described by a symbolic dynamical system with $(r+1)$ states. The first state corresponds to the position of a trajectory (solution) of (5.3) inside U, i.e., in the "background" of the field, while the other r states correspond to the position of this trajectory close to one of r points of the original homoclinic trajectory.

The corresponding topological Markov chain is defined by the following $(r+1) \times (r+1)$-transition matrix

$$A = \begin{vmatrix} 1 & 1 & 0 & \cdots & 0 \\ 0 & 0 & 1 & \cdots & 0 \\ \cdots & \cdots & \cdots & \cdots & \cdots \\ 0 & 0 & 0 & \cdots & 1 \\ 1 & 0 & 0 & \cdots & 0 \end{vmatrix} \tag{5.5}$$

Denote by Λ the set of all points of all trajectories of (5.3) that belong to $U \cup U_1$. If ϵ_1 and ϵ are small enough then all trajectories in Λ are stable with respect to the LDS (5.1) (Afraimovich *et al.*, 1994; Afraimovich & Nekorkin, 1994).

The following formula was proven by Afraimovich & Bunimovich (1995):

$$\frac{1}{n}\ln\left(1+\frac{1}{\exp(h)-1}\right) = C_1 + \frac{C_2}{\ln\left(1+\frac{\lambda_0}{2}+\sqrt{\left(1+\frac{\lambda_0}{2}\right)^2-1}\right)}, \qquad (5.6)$$

where h is the topological entropy of the topological Markov chain (5.5) and C_1, C_2 are some positive constants.

The topological entropy h is the natural characteristic of the spatial complexity of the considered class of solutions, while λ_0 is proportional to the Lyapunov exponent of the local system. Therefore (5.6) gives the relation between the local production of entropy and the spatial complexity of solutions (spatial frozen structures) of the LDS of reaction-diffusion type (5.1).

Numerical experiments confirmed the relevance of the relation (5.6) for some one-dimensional LDS of reaction-diffusion type (Giberti, 1996). A relation analogous to (5.6) for higher-dimensional LDS is not yet known. However, estimations of the topological entropy from above and from below for some d-dimensional, $d > 1$, LDS of the reaction-diffusion type have been obtained by Afraimovich & Bunimovich (1995).

Acknowledgement. This work was partially supported by U.S. National Science Foundation grant #DMS93-03769.

References

Afraimovich, V.S. & Bunimovich, L.A. (1993) Simplest structures in coupled map lattices and their stabilities, *Random and Computational Dynamics* **1**, pp. 423–444.

Afraimovich, V.S. & Bunimovich, L.A. (1995) Density of defects and spatial entropy in extended systems, *Physica D* **80**, pp. 277–288.

Afraimovich, V.A. & Chow, S.-N.(1995) Spatial chaos and homoclinic points of $\mathbb{Z}^d$-action in lattice dynamical systems, *Japan J. Indust. Appl. Math.* **12** (1995) 1–17.

Afraimovich, V.S., Chow, S.-N. & Shen, W. (1996) Hyperbolic homoclinic points of $\mathbb{Z}^d$-actions in lattice dynamical system, *Int. J. of Bifurcations and Chaos* **6**, pp. 1059–1075.

Afraimovich, V.S., Glebsky, L.Y. & Nekorkin, V. I. (1994) Stability of stationary states and topological spatial chaos in lattice dynamical systems, *Random and Computational Dynamics* **2**, pp. 287–303.

Afraimovich, V.S. & Nekorkin, V.I. (1994) Chaos of travelling waves in a discrete chain of diffusively coupled maps, *Int. J. of Bifurcations and Chaos* **4**, pp. 631–637.

Afraimovich, V.S. & Pesin, Ya.B. (1993a) Traveling waves in lattice models of multi-dimensional and multi-component media, I. General Hyperbolic Properties, *Nonlinearity* **6**, pp. 429–455,

Afraimovich, V.S. & Pesin, Ya.B. (1993b) Traveling waves in lattice models of multi-dimensional and multi-component media, II. Ergodic Properties and Dimension, *Chaos* **3**, pp. 233–241.

Aubry, S. (1992) The concept of anti-integrability: definitions, theorems and applications, in McGehee, R. & Meyer, K.E. (eds.) *Twist Mappings and Their Applications*, pp. 7–54. Springer Verlag.

Boldrighini, C., Bunimovich, L.A., Cosimi, G., Frigio, A. & Pellegrinotti, A. (1995) Ising-type transitions in coupled map lattices, *J. Stat. Phys.* **80**, pp. 1185-1205.

Bourgain, J. (1994a) On the Cauchy and invariant measure problem for the periodic Zakhavov system, *Duke Math. J.* **76**, pp. 175–202.

Bourgain, J. (1994b) Periodic nonlinear Schrödinger equations and invariant measures, *Comm. Math. Phys.* **166** (1994) 1–26.

Bourgain, J. (1996) Invariant measures for the 2D-defocusing nonlinear Schrödinger equation, *Comm. Math. Phys.* **176**, pp. 421–445.

Bowen, R. (1975) *Equilibrium states and the ergodic theory of Anosov diffeomorphisms.* Lect. Notes in Math. Vol. 470, Springer Verlag, NY.

Bricmont, J. & Kupiainen, A. (1995) Coupled analytic maps, *Nonlinearity* **8**, pp. 379–396.

Bricmont, J. & Kupiainen, A. (1996) High temperature expansions and dynamical systems, *Comm. Math. Phys.* **178**, pp. 703–732.

Bricmont, J. & Kupiainen, A. (1997) Infinite dimensional SRB measures, *Physica D* **103**, pp. 18–33.

Bunimovich, L.A. (1995) Coupled map lattices: one step forward and two steps back, *Physica D* **86**, pp. 248–255.

Bunimovich, L.A. & Carlen, E.A. (1995) On the problem of stability in lattice dynamical systems, *J. Diff. Eqns.* **123**, pp. 213–229.

Bunimovich, L.A. & Carlen, E.A. (1996) On stability of structures in lattice dynamical systems (in progress).

Bunimovich, L.A., Franceschini, V., Giberti, C. & Vernia, C. (1997) On stability of structures and patterns in extended system, *Physica D* **103**, pp. 412–418.

Bunimovich, L.A., Lambert, A. & Lima, R. (1990) The emergence of coherent structures in coupled map lattices, *J. Stat. Phys.* **61**, pp. 253–262.

Bunimovich, L.A., Livi, R., Martinez-Mekler, G. & Ruffo, S. (1992) Coupled trivial maps, *Chaos* **2**.

Bunimovich, L.A. & Sinai, Ya.G. (1988) Space-time chaos in coupled map lattices, *Nonlinearity* **1**, pp. 491–518.

Bunimovich, L.A. & Sinai, Ya.G. (1993) Statistical mechanics of coupled map lattices, in *Theory and Applications of Coupled Map Lattices.* J. Wiley.

Bunimovich, L.A. & Venkatagiri, S. (1996) Onset of chaos in coupled map lattices via peak-crossing bifurcations, *Nonlinearity* **9**, pp. 1281–1298.

Chow, S.-N. & Mallet-Paret, J. (1995) Pattern formation and spatial chaos in lattice dynamical systems I, *IEEE Trans. Circuit Syst.* **42** pp. 1–6; II. ibid. pp. 1–5.

Chow, S.-N. & Shen, W. (1995) Stability and bifurcation of travelling wave solutions in coupled map lattices, *Dyn. Systems and Appl.* **4**, pp. 1–25.

Coullet, P. & Huerre, P., Eds. (1990) *New Trends in Nonlinear Dynamics and Pattern-Forming Phenomena.* Plenum, NY.

Crutchfield, J. & Kaneko, K. (1987) Phenomenology of the space-time chaos, pp. 272–353 in Lin, H.-B. (ed.) *Directions in Chaos.* World Scientific.

Dobrushin, R.L. (1968) The problem of uniqueness of a Gibbsian random field and the problem of phase transitions, *Funct. Anal. Appl.* **2**, pp. 302–312.

Egolf, D. & Greenside, H. (1995) Preprint.

Giberti, C. (1996) Calculation of density of defects in lattice systems, *Random and Computational Dynamics* (to be published).

Giberti, C. & Vernia, C. (1993) On the presence of normally attracting manifolds containing periodic or quasiperiodic orbits in coupled map lattices, *Int. J. Bifurcations and Chaos* **3**, pp. 103–1514.

Giberti, C. & Vernia, C. (1994) Periodic behavior in 1D and 2D coupled map lattices of small size, *Chaos* **4**, pp. 651–664.

Gundlach, V.M. & Rand, D.A. (1993a) Spatial-temporal chaos: 1. Hyperbolicity, structural stability, spatial-temporal shadowing and symbolic dynamics, *Nonlinearity* **6**, pp. 165–200.

Gundlach, V.M. & Rand, D.A. (1993b) Spatial-temporal chaos: 2. Unique Gibbs states

for higher-dimensional symbolic systems, *Nonlinearity* **6**, pp. 201–214.
Gundlach, V.M. & Rand, D.A. (1993c) Spatial-temporal chaos: 3. Natural spatial-temporal measures for coupled circle map lattices, *Nonlinearity* **6**, pp. 215–230.
Jiang, M. (1995) Equilibrium states for lattice models of hyperbolic type, *Nonlinearity* **8**, pp. 631–659. See also, PhD thesis Penn State Univ., 1995.
Jiang, M. & Mazel, A. (1996) Uniqueness and exponential decay of correlations for some 2-dimensional spin-lattice systems, *J. Stat. Phys.* **82**, pp. 797-821.
Kaneko, K., Ed. (1993a) *Theory and Applications of Coupled Map Lattices.* J. Wiley.
Kaneko, K., Ed. (1993b) *Focus Issue on Coupled Map Lattices, Chaos* **2**.
Katok, A.B. & Hasselblatt, B. (1995) *Introduction to the Modern Theory of Dynamical Systems.* Cambridge Univ. Press.
Keller, G. & Künzle, M. (1992) Transfer operators for coupled map lattices, *Er. Th. and Dyn. Syst.* **12**, pp. 297–318.
Lebowitz, J., Rose, R. & Speer, E. (1986) Statistical mechanics of the nonlinear Schrödinger equation, *J. Stat. Phys.* **60**, pp. 657–687.
Lesieur, M., Ed. (1989) *Organized Structures and Turbulence.* Kluver, Dordrecht.
MacKay, R.S. (1996) Dynamics of networks: features which persist from the uncoupled limit, pp. 81–104 in van Strien, S.J. & Verduyn Lunel, S.M. (eds.) *Stochastic and spatial structures of dynamical systems.* North Holland.
Malyshev, V.A. & Minlos, R.A. (1991) *Gibbs Random Fields, Clusters Expansions.* Kluver, Dordrecht.
McKean, H. & Vaninski, K. (1994) Statistical mechanics of nonlinear wave equations and Brownian motion with restoring drift, *Comm. Math. Phys.* **160**, pp. 615–630.
Miller, J. & Huse, D.A. (1993) Macroscopic equilibrium from microscopic irreversibility in a chaotic coupled-map lattice, *Phys. Rev. E.* **48**, pp. 2528–2535.
Pesin, Ya.B. & Sinai Ya.G. (1991) Space-time chaos in chains of weakly coupled hyperbolic maps, in Sinai, Ya. G. (ed.) *Advances in Soviet Mathematics*, v. 3, Harwood.
Ruelle, D. (1978) *Thermodynamic Formalism.* Addison-Wesley.
Shen, W. (1995) Lifted lattices, hyperbolic structures and topological disorders of coupled map lattices, Preprint, Auburn University.
Simon, B. (1994) *The Statistical Mechanics of Lattice Gases*, Vol. 1. Princeton Univ. Press.
Sinai, Ya.G. (1972) Gibbs measures in ergodic theory, *Russ. Math. Surveys* **27**, pp. 21–64.
Sinai, Ya.G. (1994) *Topics in Ergodic Theory.* Princeton Univ. Press.
Sinai, Ya.G. (1996) A remark concerning the thermodynamical limit of Lyapunov spectrum, *Int. J. of Bifurcations and Chaos* **6**, pp. 1137–1142.
Temam, R. (1988) *Infinite Dimensional Dynamical Systems in Mechanics and Physics.* Springer, NY.
Volevich, D.L. (1991) Kinetics of coupled map lattices, *Nonlinearity* **4**, pp. 37–45.
Volevich, D.L. (1993) The Sinai-Bowen-Ruelle measure for a multidimensional lattice of interacting hyperbolic mappings, *Russ. Acad. Dokl. Math.* **47**, pp. 117–121.
Volevich, D.L. (1994) Construction of an analogue of Sinai-Bowen-Ruelle measure for a multidimensional lattice of interacting hyperbolic mappings, *Russ. Acad. Math. Sbornik* **79**, pp. 347–363.

ATTRACTORS AND DYNAMICS IN PARTIAL DIFFERENTIAL EQUATIONS

JACK K. HALE
Center for Dynamical Systems and Nonlinear Dynamics
School of Mathematics
Georgia Institute of Technology
Atlanta, GA 30332
U.S.A.
email: hale@math.gatech.edu

Abstract.

In these lectures, we first give an introduction to the theory of global attractors for abstract semigroups on a Banach space. In the next section, we illustrate these concepts by discussing a reaction diffusion equation. Also, we give a survey of some of the recent results that have been obtained concerning the flow on the global attractor: connecting orbits, transversality properties, slow motion manifolds, and convergence properties. In the third section, we consider a damped hyperbolic equation, concentrating on the additional complexities that arise because the solution operator is not a compact operator. In the final section we discuss the dependence of global attractors on the parameters which occur in the governing equation.

1. Basic Concepts

If we consider an evolutionary equation in a Banach space X and assume that the vector field does not depend upon time, then the solution through a point $x \in X$ at time 0 may be represented as $T(t)x$. If we assume also that each solution is defined for all $t \geq 0$ and that there a unique solution through each point, then the family of operators $T(t), t \geq 0$, satisfies the following properties:

$$T(0) = I, \quad T(t + \tau) = T(t)T(\tau). \tag{1.1}$$

J. C. Robinson and P. A. Glendinning (eds.),
From Finite to Infinite Dimensional Dynamical Systems, 85–112.

A family of transformations $T(t), t \geq 0$, is said to be a *semigroup* of transformations or a *dynamical system* if it satisfies (1.1) and $T(t)x$ is continuous in t, x.

In many situations, the dynamics of a system is defined by the properties of the iterate of a map T on X rather than by an evolutionary equation. In this case, the time is discrete and represented by the integers. We could present both of these cases together by taking a family of transformations indexed by a parameter in a group. However, we present only the continuous case for simplicity.

If the semigroup $T(t)$ is defined by the solutions of an ordinary differential equation in $\mathbb{R}^n$, then the compactness of the unit ball in $\mathbb{R}^n$ implies that $T(t)$ has the additional property that it is a compact map; that is, takes bounded sets into sets whose closure is compact. As we shall see, certain types of very important infinite dimensional systems enjoy this same property even though the unit ball is not compact. More specifically, this will be the case for reaction diffusion equations and retarded differential difference equations with a finite delay.

We introduce now some of the fundamental concepts of dynamical systems. For any point $x \in X$, we define the *positive orbit through* x as the set $\gamma^+(x) = \cup_{t \geq 0} T(t)x$. For any set $B \subset X$, we define the *positive orbit through* B as the set $\gamma^+(B) = \cup_{x \in B} \gamma^+(x)$. A set $J \subset X$ is said to be *invariant* if $T(t)J = J$ for $t \geq 0$. Since $T(t)$ maps an invariant set J onto itself, it follows that, for any point $x \in J$ and any $t \geq 0$, there is a $y \in J$ such that $T(t)y = x$. As a consequence, we can define $T(-t)x = y$ and obtain a negative orbit through x. This corresponds to defining a *backward continuation of an orbit through* x if x belongs to an invariant set J. When we can define a negative orbit through x, we refer to the set $\gamma(x) = \cup_{t \in \mathbb{R}} T(t)x$ as a *globally defined orbit.*

Of fundamental importance in dynamical systems is the *maximal bounded invariant set* $\mathcal{A}$. If this set is bounded, then it consists of the set of globally defined bounded orbits.

For $x \in X$, we define $\omega(x)$, the *ω-limit set of x* or the *ω-limit set of the orbit through x*, as

$$\omega(x) = \cap_{\tau \geq 0} \mathrm{Cl}\gamma^+(T(\tau)x).$$

This is equivalent to saying that $y \in \omega(x)$ if and only if there is is a sequence $t_n \to \infty$ such that $T(t_n)x \to y$ as $n \to \infty$. In the same way, for any set $B \subset X$, we define $\omega(B)$, the *ω-limit set of B* or the *ω-limit set of the orbit through B*, as

$$\omega(B) = \cap_{\tau \geq 0} \mathrm{Cl}\gamma^+(T(\tau)B).$$

This is equivalent to saying that $y \in \omega(B)$ if and only if there are sequences $t_n \to \infty$, $x_n \in B$, such that $T(t_n)x \to y$ as $n \to \infty$.

We remark that $\omega(B)$ may not be $\cup_{x\in B}\omega(x)$. In fact, consider the simple example, $\dot{x} = x - x^3$ and the set B to be the interval $[-2, 2]$. Then $\omega(B) = [-1, 1]$ and $\cup_{x\in B}\omega(x) = \{0, \pm 1\}$.

We say that a set A *attracts* a set B under the flow defined by the semigroup $T(t)$ if $\lim_{t\to\infty} \operatorname{dist}(T(t)B, A) = 0$, where

$$\operatorname{dist}(B, A) = \sup_{x\in B} \operatorname{dist}(x, A).$$

A basic but very elementary result is the following

Lemma 1.1. *If $B \subset X$ and $\mathrm{Cl}\,\gamma^+(B)$ is compact, then $\omega(B)$ is compact, invariant, and $\omega(B)$ attracts B.*

In the case where $X = \mathbb{R}^n$, the hypotheses of Lemma 1.1 are satisfied if we only assume that $\gamma^+(B)$ is bounded since the closure of bounded sets in $\mathbb{R}^n$ are compact.

If there is a $t_0 \geq 0$ such $T(t)$ is compact for $t \geq t_0$, then the hypotheses of Lemma 1.1 also are satisfied if we assume only that $\gamma^+(B)$ is bounded. In this case, the unit ball in X is not compact if X is infinite dimensional, but this inconvenience is overcome by the fact that the mapping $T(t)$ is compact for $t \geq t_0$. Semigroups which are compact for $t \geq t_0$ occur in many applications including, for example, reaction diffusion equations and retarded differential difference equations.

There are many other applications for which the semigroup is not compact for any $t \geq 0$. As we will see, this is the case for a wave equation and yet, if there is dissipation in the equation, the conclusions of Lemma 1.1 hold. At this point, we give a general class of semigroups for which Lemma 1.1 holds and make the applications later.

Motivated by the fact that $T(t)\gamma^+(B) \subset \gamma^+(B)$ for all $t \geq 0$ and that $\omega(B)$ attracts B is equivalent to $\omega(B)$ attracts $\gamma^+(B)$, we make the following definition. We say that the semigroup $T(t), t \geq 0$ is *asymptotically smooth* if, for any bounded set B in X with $T(t)B \subset B$ for $t \geq 0$, there exists a compact set J in the closure of B such that J attracts B. We notice that, if $T(t)$ is compact for $t \geq t_0$, then $T(t), t \geq 0$, is asymptotically smooth. Also, if

$$T(t) = S(t) + U(t), \tag{1.2}$$

where $U(t)$ is compact for $t \geq 0$ and there are continuous functions $k(r) > 0$, $\alpha(r) > 0$, defined for $r \geq 0$, such that

$$|S(t)x| \leq k(r)e^{-\alpha(r)t}, \quad t \geq 0, \quad |x| \leq r, \tag{1.3}$$

then $T(t)$ is asymptotically smooth. For each $t > 0$, the operator $T(t)$ may not be compact, but as $t \to \infty$ it becomes compact. The semigroup $T(t)$

also is asymptotically smooth if there are positive constants k, α and a precompact pseudo metric $\rho(t, \cdot, \cdot)$ such that

$$|T(t)x - T(t)y| \le ke^{-\alpha t}|x - y| + \rho(t, x, y)$$

for all $t \ge 0$, $x, y \in X$ (see Hale, 1988).

If $T(t)$ is a linear semigroup, then there is a decomposition $T(t) = S(t) + U(t)$ where $U(t)$ is compact for $t > 0$ and the spectral radius of $S(t)$ is the same as the radius of the essential spectrum of $T(t)$. Therefore, $T(t)$ is asymptotically smooth if and only if, for each $t > 0$, the radius of the essential spectrum of $T(t)$ is less than one (see Deimling, 1985).

Lemma 1.2. *If $T(t)$ is asymptotically smooth and $\gamma^+(B)$ is bounded, then $\omega(B)$ is compact, invariant, and $\omega(B)$ attracts B.*

The next important concepts deal with attractors. A set $\mathcal{A}$ is said to be a *minimal global attractor* if $\mathcal{A}$ is invariant, attracts bounded sets and is minimal with respect to this property (Ladyzenskaya (1987)). If the minimal global attractor is compact, we will refer to the compact minimal global attractor as the *global attractor.*

Lemma 1.3. *If $T(t)$ is asymptotically smooth and, for each bounded set B, $\gamma^+(B)$ is bounded, then there exists a minimal global attractor $\mathcal{A}$, which is locally compact and is given by*

$$\mathcal{A} = \cup_{B \subset X} \omega(B).$$

For the example $\dot{x} = -\epsilon x, \dot{y} = -y$, the minimal global attractor is the x-axis for $\epsilon = 0$ and the global attractor is the origin for $\epsilon > 0$. In this example, a small dissipation in the x-direction turned the minimal global attractor into a global attractor. This is a general phenomenon for the case where $T(t)$ is compact for some $t > 0$.

We say that $T(t)$ is *point dissipative* if there is a bounded set B in X such that, for any $x \in X$, there is a $t_0 = t_0(x, B)$ such that $T(t)x \in B$ for $t \ge t_0$.

Theorem 1.1. *If $T(t)$ is point dissipative and there is an $r \ge 0$ such that $T(t)$ is compact for $t \ge r$, then there exists the global attractor $\mathcal{A}$.*

Theorem 1.1 can be useful in many situations; in particular, it has applications to ordinary differential equations, reaction diffusion systems and retarded functional differential equations with finite delay.

For the case where $T(t)$ is asymptotically smooth, we need an additional condition to ensure the existence of the global attractor.

Theorem 1.2. *If $T(t)$ is asymptotically smooth, $\gamma^+(B)$ is bounded if B is bounded, and $T(t)$ is point dissipative, then there exists the global attractor $\mathcal{A}$.*

Theorem 1.2 can be very useful in the discussion of partial differential equations for which the family of maps $T(t)$ is a group; for example, the damped wave equation, and the Fitzhugh-Nagumo equation. It also is appropriate for some neutral functional differential equations and retarded equations with infinite delay.

It is difficult to trace the earliest papers which contained discussions involving concepts related to global attractors. The idea is very old and the interested reader can consult Babin & Vishik (1991), Hale (1988), Hale *et al.* (1984), Hale & Verduyn-Lunel (1993), Haraux (1989), Ladyzenskaya (1987), Temam (1988) and the references therein.

We remark that we can have $\gamma^+(B)$ bounded for B bounded and $T(t)$ point dissipative and yet the global attractor may not exist. In fact, consider the example

$$\frac{d}{dt}[x(t) - x(t-1)] = -x(t)$$

which can arise from the discussion of idealized lossless transmission lines. If we consider an initial value φ in the space $C([-1,0],\mathbb{R})$, then there is a unique continuous solution $x(t)$ for $t \geq -1$, $x(t) = \varphi(t)$, $t \in [-1,0]$. If we define $(T(t)\varphi)(\theta) = x(t+\theta)$, $\theta \in [-1,0]$, $t \geq 0$, then $T(t)$ is a semigroup on $C([-1,0],\mathbb{R})$. Solutions of the form $\exp(\lambda t)$ yield the eigenvalues as solutions of the equation $\lambda(1-e^{-\lambda}) + 1 = 0$. The eigenvalues are simple and have negative real parts. This is enough to imply that orbits of bounded sets are bounded and solutions approach zero as $t \to \infty$. Also, the set of globally defined solutions is the origin. If there was a global attractor, then it would be the origin. But, since the equation is linear, this would mean that the origin is stable and that $|T(t)| \to 0$ as $t \to \infty$ exponentially, which cannot be the case.

Another example is the heat equation $u_t = \Delta u$ on $\mathbb{R}$ in the space $C_0(\mathbb{R},\mathbb{R})$, the space of continuous functions $\varphi(x)$ on $\mathbb{R}$ which approach zero as $|x| \to \infty$. It is not difficult to see that each solution approaches zero as $t \to \infty$ and yet the approach is not exponential since the essential spectral radius of the semigroup generated by this equation is one for all $t \geq 0$.

We will see another example of this type when we consider a locally damped wave equation.

If a dynamical system has a global attractor, it is important to have some estimate of its size. To be precise about the size, we must recall some definitions of dimension.

Let K be a topological space. We say that K is finite dimensional if there exists an integer n such that, for every open covering A of K, there exists another open covering A' refining A such that every point of K belongs to at most $n+1$ sets of A'. In this case, the *dimension* of K, dim K, is defined as the minimum n satisfying this property. The definition implies that $\dim \mathbb{R}^n = n$. Also, if K is a compact finite dimensional space, then the Whitney embedding theorem implies that K is homeomorphic to a subset of $\mathbb{R}^n$ with $n = 2\dim K + 1$.

If K is a metric space, its *Hausdorff dimension*, $\dim_H(K)$, is defined as follows: for any $\alpha > 0, \epsilon > 0$, let

$$\mu_\epsilon^\alpha(K) = \inf \sum_i \epsilon_i^\alpha$$

where the infimum is taken over all coverings $B_{\epsilon_i}(x_i)$, $i = 1, 2, \ldots$, of k with $\epsilon_i < \epsilon$ for all i, where $B_{\epsilon_i}(x_i) = \{x : d(x, x_i) < \epsilon_i\}$. Let $\mu^\alpha(K) = \lim_{\epsilon \to 0} \mu_\epsilon^\alpha(K)$ and define the Hausdorff dimension of K as

$$\dim_H(K) \equiv \inf\{\alpha : \mu^\alpha(K) = 0\}.$$

To define another measure of the size of a metric space K, let $n(\epsilon, K)$ be the minimum number of open balls of radius ϵ needed to cover K. Define the *limit capacity* or *box dimension* $c(K)$ of K as

$$c(K) = \limsup_{\epsilon \to 0} \frac{\log n(\epsilon, K)}{\log(1/\epsilon)}.$$

In other words, $c(K)$ is the minimum real number such that, for every $\sigma > 0$, there is a $\delta > 0$ such that

$$n(\epsilon, K) \leq \epsilon^{-c(K)-\sigma} \text{ if } 0 < \epsilon < \delta.$$

We always have $\dim_H(K) \leq c(K)$.

An important result in the applications is

Theorem 1.3. *Suppose that X is a Banach space, $T(t) : X \to X$ satisfies (1.2), (1.3), is bounded dissipative and positive orbits of bounded sets are bounded. Then the global attractor $\mathcal{A}$ has the following properties:*

1. *$c(\mathcal{A}) < \infty$.*
2. *If $d = 2c(\mathcal{A}) + 1$ and S is any linear subspace of X with $\dim S \geq d$, then there is a residual set Π of the space of all continuous projections P of X onto S (taken with the uniform operator topology) such that $P|\mathcal{A}$ is one-to-one for every $P \in \Pi$.*

For compact maps and X a Hilbert space, Mallet-Paret (1976) proved that $\dim_H(\mathcal{A}) < \infty$. This result as stated is due to Mañé (1981). Having the capacity finite implies the Hausdorff dimension finite. The second part of the theorem says that the global attractor can be unravelled in a residual set of directions onto a finite dimensional subspace of large dimension.

2. A Scalar Reaction Diffusion Equation

Suppose that Ω is a bounded domain in $\mathbb{R}^N$ with smooth boundary and consider the equation

$$\begin{aligned} u_t - \Delta u &= f(u) \quad \text{in } \Omega, \\ \frac{\partial u}{\partial n} &= 0 \quad \text{on } \partial\Omega, \end{aligned} \tag{2.1}$$

and initial condition $u|_{t=0} = u_0 \in H^1(\Omega)$.

To obtain solutions of (2.1) in the space $H^1(\Omega)$, we suppose that $f \in C^2(\mathbb{R}, \mathbb{R})$ and also satisfies the following growth conditions for constants c, γ:

$$\begin{aligned} |f'(u)| &\leq ce^{|u|^\gamma}, \ 0 \leq \gamma < 2, \ n = 2, \\ |f'(u)| &\leq c(1 + |u|^{\gamma+1}), \ 0 \leq \gamma \leq 1, n = 3. \end{aligned} \tag{2.2}$$

Under condition (2.2), it is possible to prove the local existence and uniqueness of the solution of (2.1) with initial data $\varphi \in H^1(\Omega)$, see for example Henry (1981). To obtain the global existence, we need to obtain a uniform bound on the orbit and this will be satisfied if we impose the following dissipative condition for some positive constant λ_1:

$$\limsup_{|u|\to\infty} \frac{f(u)}{u} \leq -\lambda_1 \tag{2.3}$$

Actually, we obtain global existence by only assuming $\lambda_1 = 0$. The stronger condition will give us a global attractor.

The a priori bounds on solutions are obtained by using the natural energy function

$$V(\varphi) = \int_\Omega [\frac{1}{2}|\nabla\varphi|^2 - F(\varphi)]dx,$$

where $F(u) = \int_0^u f$. In fact, we easily obtain by differentiation and an integration by parts

$$\dot{V}(u(t)) = -\int_\Omega u_t^2 \leq 0, \tag{2.4}$$

or $V(u(t)) \leq V(\varphi)$ for $t \geq 0$. The dissipative condition (2.3) implies that $V(\varphi) \to \infty$ as $|\varphi| \to \infty$ and we obtain an a priori bound on the solutions. This implies global existence as well as the fact that positive orbits of bounded sets are bounded. If $T(t)\varphi$ is defined to be the solution through φ, then $T(t)$, $t \geq 0$, is a C^1-semigroup.

Furthermore, $T(t)$ is a compact map for any $t > 0$. This fact together with the boundedness of $\gamma^+(\varphi)$ implies that $\omega(\varphi)$ is a compact, connected invariant set. Relation (2.4) then implies that $\omega(\varphi) \subset E$, where E is the set of equilibrium points.

The dissipative condition (2.3) implies also that the set E is bounded. Since $\omega(\varphi)$ attracts φ and $\omega(\varphi) \subset E$, we have point dissipativeness.

From the above discussion and Theorem 1.1, there is a global attractor $\mathcal{A}$ for (2.1). Since $\mathcal{A}$ is invariant, it follows that $\mathcal{A} = W^u(E)$, the unstable set of the set E of equilibria.

We say that an equilibrium point φ is hyperbolic if the spectrum $\sigma(\Delta + f'(\varphi))$ of $\Delta + f'(\varphi)$ does not intersect the imaginary axis in the complex plane. Since this operator is self adjoint, this is equivalent to saying that there is no zero eigenvalue. For an hyperbolic equilibrium point φ, the spectrum of the linear operator $L = \Delta + f'(\varphi)$ leads to a natural decomposition of $H^1(\Omega)$ into two subspaces W_L^u, W_L^s which are invariant under L and $L|W_L^u$ has positive spectrum (which is finite) and $L|W_L^s$ has negative spectrum bounded away from zero. For the nonlinear equation, we can define the unstable set $W^u(\varphi)$ of φ as those points whose orbits approach φ as $t \to -\infty$ and the stable set $W^s(\varphi)$ as those points whose orbits approach φ as $t \to \infty$. These are C^1 manifolds which are tangent at φ respectively to W_L^u and W_L^s.

If all of the equilibrium points are hyperbolic, then

$$\mathcal{A} = \cup_{\varphi \in E} W^u(\varphi). \tag{2.5}$$

System (2.1) belongs to the class of systems, called *gradient systems*, which are those for which there is a Lyapunov function (playing the role of $V(\varphi)$ above) and for which the ω-limit set of any orbit is in the set of equilibrium points. If a global attractor exists for such a system and the equilibrium points are hyperbolic, then (2.5) holds.

Understanding more about the flow on the attractor $\mathcal{A}$ involves determining the manner in which the equilibrium points are connected by orbits. We also want to know the sensitivity of the flow to perturbations in the equation (structural stability). So far, most of the detailed properties of the flow are known in only one space variable.

We now restrict attention to one space variable:

$$\begin{aligned} u_t - u_{xx} &= f(u) \quad \text{in } (-1,1), \\ u_x &= 0 \quad \text{at } x = -1, x = 1, \end{aligned} \tag{2.6}$$

We remark that we could have other separated boundary conditions at $x = -1, x = 1$ as well as a function $f(x, u, u_x)$. The system still would be gradient.

A gradient system is *Morse-Smale* if it has only a finite number of equilibrium points and the stable and unstable manifolds are transversal.

Theorem 2.1. *If (2.6) is dissipative, then it is Morse-Smale if and only if the equilibrium points are hyperbolic.*

This very surprising result of Angenent (1986) and Henry (1985) implies that the stable and unstable manifolds of hyperbolic equilibrium points *always intersect transversally.* In the class of ordinary differential equations with general vector fields, the only ones for which such a strong property is true are the scalar equations.

Let us give an idea of the proof of this result which relies heavily upon the Sturm-Liouville theory and the Jordan curve theorem. The first important remark is the following

Lemma 2.1. *If φ, ψ are hyperbolic equilibrium points of (2.6), and there is a ξ such that $\alpha(\xi) = \varphi$, $\omega(\xi) = \psi$, then dim $W^u(\varphi) >$ dim $W^u(\psi)$.*

Let us mention the ingredients of the proof of Lemma 2.1. If we designate the eigenpairs of $\partial^2/\partial^2 x + f'(\varphi)$ by (φ_j, λ_j), $\lambda_1 > \lambda_2 > \ldots$, then the tangent space $TW^u(\varphi)$ to $W^u(\varphi)$ at φ is given by $TW^u(\varphi) = \operatorname{span}\{\varphi_j : j - 1, 2, \ldots, \lambda_{i(\psi)}\}$, where $i(\varphi) = \dim W^u(\varphi)$. The Sturm-Liouville theory implies that the number of zeros $z(\varphi_j)$ in $(-1, 1)$ is $j - 1$. In the same way, if we designate the eigenpairs of $\partial^2/\partial^2 x + f'(\psi)$ by (ψ_j, μ_j), $\mu_1 > \mu_2 > \ldots$, then the tangent space $TW^s(\psi)$ to $W^s(\psi)$ at ψ is the orthogonal complement to $\operatorname{span}\{\psi_j : j = 1, 2, \ldots, \mu_{i(\psi)}\}$, where $i(\psi) = \dim W^u(\psi)$. Also, the number of zeros $z(\psi_j)$ in $(-1, 1)$ is $j - 1$. Therefore, we have that $\bar{\varphi} \in TW^u(\varphi)$ implies that $z(\bar{\varphi}) \le i(\varphi) - 1$ and $\bar{\psi} \in TW^s(\psi)$ implies that $z(\bar{\psi}) \ge i(\psi)$ (this latter property uses the nontrivial fact that no solution of the linearized parabolic problem can approach zero faster than any exponential). If $\eta(t)$ is the solution of (2.6) through ξ, then $\eta(t) \to \varphi$ along $TW^u(\varphi)$ as $t \to -\infty$ and $\eta(t) \to \psi$ along $TW^s(\psi)$ as $t \to \infty$ with the approach being exponential. Since $\eta'(t)$ satisfies the linear equation about η

$$v_t = v_{xx} + f'(\eta(t))v,$$

it is reasonable to expect that (and this is what can be proved) $z(\eta'(t)) \le i(\varphi) - 1$ for t close to $-\infty$ and $z(\eta'(t)) \ge i(\psi)$ for t close to ∞.

The next step in the proof is the observation that, for any solution $v(t)$ of the above linear variational equation, $z(v(t))$ is a non-increasing function of t. The number of zeros of a function in $(-1, 1)$ serves as a *discrete Lyapunov function.* The proof of this fact uses the maximum principal and the Jordan curve theorem (Nickel, 1962; Matano, 1982). If we use this result and the above property of η', we obtain Lemma 2.1.

To prove Theorem 2.1, one supposes that the manifolds are not transversal, uses the characterization of $TW^s(\psi)$ in terms of the adjoint of the linear variational equation around ψ, and argues as in the proof of Lemma 2.1 for the adjoint equation to obtain a contradiction to Lemma 2.1.

We remark that the discussion of the spatial discretization of (2.6) in the light of Theorem 2.1 has led to some very interesting results. This spatial discretization leads to a system of ordinary differential equations $\dot{U} = F(U)$, where the Jacobian of F is a *tri-diagonal Jacobi matrix*; that is, the Jacobian is tri-diagonal with the off diagonal elements positive.

Theorem 2.2. *For an ordinary differential equation with tri-diagonal Jacobian, the stable and unstable manifolds of hyperbolic equilibrium points always intersect transversally.*

This result is due to Fusco & Oliva (1988). The proof also uses a discrete Lyapunov function defined by the number of sign changes in the coordinates of U.

Similar results are known for the time discretization of (2.6) as well as the time and space discretization (Oliva *et al.*, 1993).

It is natural to enquire if the system (2.1) in several space variables also satisfies the property that stable and unstable manifolds of hyperbolic equilibrium points intersect transversally. This is not known for (2.1). However, if we allow the function $f = f(x, u)$ to depend upon x, then Poláčik (1994) has given an example where we do not have transversality holding. Also, in this same class of functions $f(x, u)$, Brunovský & Poláčik (1997) have shown that generically (2.1) is a Morse-Smale system.

Lemma 2.1 leads to an interesting Morse decomposition of the attractor $\mathcal{A}$ when all of the equilibrium points are hyperbolic. A *Morse decomposition* of the global attractor $\mathcal{A}$ is a finite ordered collection $\mathcal{A}_1 < \mathcal{A}_2 < \cdots < \mathcal{A}_M$ of disjoint compact invariant subsets of $\mathcal{A}$ (called *Morse sets*) such that, for any $\varphi \in \mathcal{A}$, there are integers $N \geq K$ such that $\alpha(\varphi) \subset \mathcal{A}_N$, $\omega(\varphi) \subset \mathcal{A}_K$ and $N = K$ implies that $\varphi \in \mathcal{A}_N$.

For (2.6), if we define $\mathcal{A}_N = \{\varphi \in E : i(\varphi) = N\}$, then there is an integer M such that these sets are empty if $N > M$. It is easy to see that the collection of sets $\{\mathcal{A}_N, N = 0, 1, 2, \ldots, M\}$ are Morse sets for a Morse decomposition. The attractor $\mathcal{A}$ is the union of the Morse sets together with all of the orbits connecting them.

This leads us naturally to the *connection problem* for (2.1). We say that $C(\varphi, \psi)$ is an orbit connecting φ to ψ if, for any point $\xi \in C(\varphi, \psi)$, we have $\alpha(\xi) = \varphi$, $\omega(\xi) = \psi$. Assuming that all of the equilibrium points are hyperbolic, an important problem is to characterize the connecting orbits. Using Lemma 2.1, Henry (1985) was able to characterize the attractor for the case when $f(u) = u - u^3$. To describe the most general result for (2.6), we use the shooting method to obtain the set E of equilibrium points; that is, we consider the initial value problem

$$u_x = v, \quad v_x = -f(u), \quad u(0) = u_0, \; v(0) = 0. \tag{2.7}$$

If $(u(x, u_0), v(x, u_0))$ is the solution of (2.7) and

$$S = \{(u(1, u_0), v(1, u_0)) : u_0 \in \mathbb{R}\},$$

then $u(\cdot, u_0) \in E$ if and only if $(u(1, u_0), v(1, u_0)) \in S \cap \{v = 0\}$.

We can now order the equilibrium points by the natural order on $\{v = 0\}$ and we also can order them along the set S by the order in which they intersect $\{v = 0\}$. This gives a permutation π of the equilibrium points (Fusco & Rocha, 1991). It is remarkable that this permutation characterizes the connections between equilibrium points. To describe the precise result, let us consider two systems (2.6) with corresponding dissipative vector fields f_0 and f_1. Let $\mathcal{A}_0$, $\mathcal{A}_1$ be the global attractors and E_0, E_1 the equilibrium sets and assume that all of the equilibrium points are hyperbolic.

Definition 2.1. The attractors $\mathcal{A}_0$, $\mathcal{A}_1$ are *connection equivalent* if there exists a bijection $\sigma : E_0 \to E_1$ such that

1. $i(\sigma(\varphi)) = i(\varphi)$ for all $\varphi \in E_0$.
2. $C(\varphi, \psi)$ exists if and only if $C(\sigma(\varphi), \sigma(\psi))$ exists.

Theorem 2.3. $\pi_0 = \pi_1$ *implies that* $\mathcal{A}_0$ *and* $\mathcal{A}_1$ *are connection equivalent.*

This result is due to Fiedler & Rocha (1996). We remark that Rocha (1994) has similar results for spatial discretization of (2.6).

We mention briefly another very interesting property of (2.6) on a long time interval $(-1/\epsilon, 1/\epsilon)$, where ϵ is small. If we rescale x to $(-1, 1)$, we obtain the equation

$$u_t = \epsilon^2 u_{xx} + f(u), \quad -1 < x < 1, \tag{2.8}$$

with homogeneous boundary conditions and $f(u) = u - u^3$. It was shown by Chafee & Infante (1974), that the only stable equilibrium points of (2.8) are the constant functions ± 1. Furthermore, for each j, two equilibrium solutions φ_j^+, φ_j^- of index j bifurcate supercritically from zero at $\epsilon_j = j^2\pi^2$, $j = 1, 2, \ldots$. In the interval $(0, \epsilon_1)$, there are three equilibrium points, $\pm 1, 0$, in the interval $(\epsilon_j, \epsilon_{j+1})$, 0 has index $j + 1$ and there are exactly $2j + 3$ equilibrium points, $\pm 1, 0, \varphi_k^\pm, k = 1, \ldots, j$.

Henry (1985) has shown (it also can be deduced from Theorem 2.2) that, for $\epsilon \in (\epsilon_j, \epsilon_{j+1})$, the attractor $\mathcal{A}_\epsilon$ is the closure of $W^u(0)$, is homeomorphic to a unit ball of dimension $j + 1$ and, for each $1 \le k \le j$, the ω-limit set of $W^u(\varphi_k^\pm)$ contains $\varphi_\ell^\mp$ for $1 \le \ell \le k$ and the points ± 1.

For most points $\varphi \in X$, the scenario for the orbit through φ is that there is a transit time to a neighborhood of $\mathcal{A}_\epsilon$, at which time it moves along a neighborhood of $W^u(\varphi)$, for some equilibrium point φ, with the speed approximately the same speed that orbits move on $W^u(\varphi)$. This continues until the orbit comes close to another unstable manifold $W^u(\psi)$ and then it moves with the speed corresponding to the attractivity properties of $W^u(\psi)$ (which may be completely different from the previous speed) and this process continues until the orbit reaches a neighborhood of the stable points ± 1.

If ϵ is very small and the orbit is computed numerically, then it is observed that this scenario is never completed. In fact, the numerical computations can stabilize (in the sense that the computer can run for days and nothing happens) and yet it is stabilizing at a point which is not an equilibrium point. What is observed is that the solution develops sharp transition layers rather quickly with the motion at first being governed by the ordinary differential equation $\dot{u} = f(u)$ (this is natural since ϵ is small). After this happens, it is expected that the transition layers would move slowly and disappear. This is what happens except that the movement of the layers is exponentially small. A rather complete explanation of this phenomenon has been given by Carr & Pego (1989 & 1990), Fusco & Hale (1989), and Fusco (1990), where they characterize the properties of the unstable manifolds of the equilibrium points.

For other slow motion phenomena in one space dimension, see Kuramura *et al.* (1992) for some systems, Alikakos *et al.* (1991), Alikakos & Fusco (1992), Bates & Xun (1994 & 1995), and Grant (1995) for the Cahn-Hilliard equation, Keller & Sternberg (1993) and Pinto (1995) for (2.6) with a tri-stable potential, and the references therein.

For the parabolic equation (2.1) in several space variables, it is interesting to discuss the types of stable equilibrium points that can occur. This is a very large subject and we mention only a few results. If the domain Ω is convex, then it was shown by Casten & Holland (1978) and Matano (1979) that the only stable equilibrium points are spatially homogeneous. This result was later extended to the case where $f = f(u, \nabla u)$ by Hess (1987).

Let us briefly describe other interesting properties of parabolic equations. Consider the system

$$u_t = u_{xx} + f(x, u, u_x), \quad 0 < x < 1, \tag{2.9}$$

with separated homogeneous boundary conditions at $x = 0$ and $x = 1$. Let X be a space on which solutions of this equation define a C^r-semigroup, $r \geq 1$.

Theorem 2.4. *For any $\varphi \in X$, there is an equilibrium point $\psi = \psi(\varphi)$ such that $\omega(\varphi) = \psi$.*

This is a remarkable result since it implies that the ω-limit set is a singleton. This property often is referred to briefly by saying that the orbit is *convergent.* There have been many proofs: Zelenyak (1968) using Lyapunov functions, Matano (1978) using the maximum principle, Hale & Raugel (1992a) using the fact that the linear variational equation around an equilibrium point has the eigenvalue zero of multiplicity at most one. The result of Hale & Raugel (1992a) holds in much more general situations since it depends only upon the multiplicity of the zero eigenvalue. Many applications are contained in Hale & Raugel (1992a).

For equation (2.1) with $f = f(x, u, \nabla u)$, it follows from the theory of monotone systems that most orbits are convergent (see Smith (1995) for precise statements and historical references). On the other hand, there exist $f(x, u, \nabla u)$ such that equation (2.1) has chaotic solutions (Poláčik, 1995; Poláčik & Rybakowski, 1994). These motions will be unstable but they do exist. We remark also that there exists a function $f(x, u)$ such that, for Ω the unit ball in $\mathbb{R}^2$, there is an orbit whose ω-limit set is a circle of equilibrium points (Poláčik & Rybakowski, 1996). Of course, this set of equilibrium points must be unstable.

Let us now turn to a non-autonomous equation in one space dimension:

$$u_t = u_{xx} + f(t, x, u, u_x), \quad 0 < x < 1, \tag{2.10}$$

with homogeneous Neumann boundary conditions. Let X be some space where the solutions are well defined and depend in a C^1-way upon the initial data. If we suppose that the function f in (2.10) is periodic in t of period ω, then we can define the Poincaré map π on X by $\pi u_0 = u(\omega, \cdot, u_0)$, where $u(t, x, u_0)$ is the solution of (2.10) with initial value u_0 at $t = 0$. Then the following result is true, where the subscript π always designates that we discussing everything with respect to the discrete dynamical system on X defined by π.

Theorem 2.5.

1. *$\omega_\pi(u_0)$ is a fixed point of π.*
2. *The stable and unstable manifolds of hyperbolic fixed points of π are transversal and, thus, if (2.10) is dissipative, then π is Morse-Smale if and only if all equilibrium points are hyperbolic.*
3. *π is gradient like and the only recurrent sets are fixed points.*

Part 1 was proved by Chen & Matano (1989) for the case where $f = f(t, u)$, and the general case is in Brunovský *et al.* (1992). Part 2 was proved by Chen *et al.* (1992) and part 3 by Chen & Poláčik (1995).

Let us now turn to the discussion of (2.10) when the function f is almost periodic in t. For any almost periodic function g, we say that a function h is in the hull $\mathcal{H}(g)$ of g if there is a sequence $\{\tau_n\} \subset \mathbb{R}$ such that $\tau_n \to \infty$, $g(t+\tau_n) \to g(t)$ as $n \to \infty$ uniformly for $t \in \mathbb{R}$.

A continuous function $g(t), t \in \mathbb{R}$ is said to be *almost automorphic* if, for any sequence $t_n \to \infty$ as $n \to \infty$, there is a subsequence (which we label the same) and a continuous function $h(t), t \in \mathbb{R}$ such that $g(t+t_n) - h(t) \to 0$, $h(t-t_n) - g(t) \to 0$ pointwise for $t \in \mathbb{R}$ (Bochner (1962)). The following remarkable result (new even for ODEs) showing that limits of bounded solutions must be almost automorphic and not necessarily almost periodic is due to Shen & Yi (1995a & b).

Theorem 2.6. *If $f(t,x,u)$ is almost automorphic in t uniformly with respect to $x \in [0,1]$ and u in bounded sets, and $u(x,t)$ is a solution of (2.10) bounded for $t \geq 0$, then there is a sequence $\{\tau_n\}$, $t_n \to \infty$ as $n \to \infty$ and an almost automorphic function $v(x,t)$ such that $u(\cdot, t+\tau_n) - v(\cdot,t) \to 0$ as $n \to \infty$ uniformly on compact sets of $\mathbb{R}$ and $v(\cdot,t)$ is a solution of (2.10) with f replaced by some function $g \in \mathcal{H}(f)$. Furthermore, there are examples of functions f and solutions u such that v is not almost periodic.*

Let us now discuss the non-autonomous case in several space dimensions. Consider the equation

$$u_t = \Delta u + f(t,x,u,\nabla u), \quad x \in \Omega, \tag{2.11}$$

where Ω is a bounded domain with smooth boundary. Poláčik & Tereščák (1991) have obtained the following result.

Theorem 2.7. *If f in (2.11) is periodic in t, then most orbits of the Poincaré map converge to periodic points of π.*

However,

Theorem 2.8. *There exists a domain Ω and a function $f(t,x,u)$ such that the Poincaré map has stable periodic points which are not fixed points (there can exist subharmonics of (2.11)).*

This result of Dancer & Hess (1994) and Takač (1991) is rather surprising in light of the result mentioned above on most orbits converging to equilibrium points when the equation is autonomous and in light of Theorem 2.5.

The following result of Hess (1987) shows that the situation is simpler if the domain Ω is convex.

Theorem 2.9. *If f in (2.11) is independent of x, periodic in t and Ω is convex, then then any periodic point of the Poincaré map which is stable must be spatially constant.*

From Theorem 2.9, any stable periodic point of π must be a solution of the scalar ODE $\dot{u} = f(t, u, 0)$. On the other hand, any periodic point of π for this equation must be a fixed point of π. Therefore, any stable periodic solution of (2.11) must be a periodic solution of the mentioned ODE and must have period the same as the period of the function f.

Shen & Yi (1995b) have an analogue of Theorem 2.7 for the case when f is almost automorphic in t.

As we have remarked before, equation (2.8) rescaled to $(-1/\epsilon, 1/\epsilon)$ gives information about solutions of a reaction diffusion equation on a long time interval. If we consider (2.6) on the interval $(-\infty, \infty)$ and suppose that $f(u) = u(u-a)(1-u)$ with $a \in (0,1)$, then it is known that there is a unique traveling wave solution $\varphi^*(x - ct)$ of the equation with wave speed $\sqrt{2}a$. In an appropriate function space, this traveling wave represents a one-dimensional manifold (the equation is translation invariant in x) and is stable. It is natural to ask if such a solution can be obtained by taking the limit of a rescaled solution of (2.8) as $\epsilon \to 0$. Fusco *et al.* (1995) have shown that this is the case if we consider a close to zero. This restriction should not be necessary but the proof given is very complicated and involves many delicate estimates. The idea for the proof is very simple and comes from the structure of the attractor. In the attractor, there is a stable one dimensional manifold given by saddle point with index one together with its unstable manifold. Fusco *et al.* (1995) exploit this fact to obtain the result.

If we change the boundary conditions in (2.6) and still maintain separated boundary conditions, then results of the above type still remain valid. If we allow periodic boundary conditions, then the structure of the flow can be different. Let us discuss this in some more detail for the specific equation

$$u_t - \epsilon^2 u_{xx} + f(u)_x = g(u) \in S^1, \tag{2.12}$$

where $\epsilon > 0$ is a parameter, S^1 is the unit circle (this corresponds to periodic boundary conditions), $g(u)$ is dissipative and $f(u)$ is a strictly convex function.

For (2.12), if we choose the initial data in $H^2(S^1)$, then it can be shown that it defines a C^0-semigroup on $H^2(S^1)$ and that there is a global attractor $\mathcal{A}_\epsilon$. The following result is due to Fiedler & Mallet-Paret (1989).

Theorem 2.10. *For (2.12), for any $\varphi \in \mathcal{A}_\epsilon$, either $\omega(\varphi)$ is a traveling wave or $\alpha(\varphi)$ and $\omega(\varphi)$ belong to the set of equilibrium points.*

If $\epsilon = 0$ in (2.12), we obtain the inhomogeneous conservation law

$$u_t + f(u)_x = g(u) \text{ in } S^1. \tag{2.13}$$

We consider (2.13) in the space $BV(S^1)$ of all functions $u \in L^1(S^1)$ such that u_x is a finite measure on S^1 and we require that solutions $u(x,t)$ satisfy the entropy condition $u(x^-, t) \geq u(x^+, t)$. Let $\mathcal{A}_0$ be the set of solutions of (2.13) which are defined for all $t \in \mathbb{R}$ and bounded. The following result is due to Fan & Hale (1994 & 1995) (see also Lyberopoulos, 1990 & 1994). In the statement of the result, the subscript L^p denotes convergence in $L^p(S^1)$.

Theorem 2.11. *Assume that the zeros $a_1, a_2, \dots a_{2m+1}$ of g are simple. Then, for any $0 < p < \infty$ and any $\varphi \in \mathcal{A}_0$, we have $\alpha_{L^p}(\varphi)$ and $\omega_{L^p}(\varphi)$ are either an equilibrium point or a traveling wave with wave speed $f'(a_{2k})$ for some k. Furthermore, $\mathcal{A}_0$ is a global attractor in $L^p(S^1)$.*

We remark that we cannot have $\mathcal{A}_0$ as a global attractor because positive orbits are not compact in this space. We obtain the compactness by working in $L^p(S^1)$ and also obtain a characterization of $\mathcal{A}_0$.

It is very natural to compare the noncompact set $\mathcal{A}_0$ of (2.13) with the limit as $\epsilon \to 0$ of the compact attractors $\mathcal{A}_\epsilon$ of (2.12). Fan & Hale (1995) come very close to proving that this limit is $\mathcal{A}_0$. In fact, they prove that any orbit in $\mathcal{A}_\epsilon$, together with its α and ω limit sets, converge almost everywhere to a similar orbit of (2.13), but were unable to show that every element in $\mathcal{A}_0$ could be approximated by a sequence of elements from $\mathcal{A}_\epsilon$. See the papers for a precise statement.

3. A Linearly Damped Hyperbolic Equation

In this section, we want to illustrate some of the additional difficulties that arise when we consider partial differential equations which do not smooth in time but which are still dissipative and there exists a global attractor. We use as a model the linearly damped wave equation

$$\begin{aligned} u_{tt} + \beta u_t - \Delta u &= f(u) \quad \text{in } \Omega \subset \mathbb{R}^n \\ u &= 0 \quad \text{on } \partial\Omega, \end{aligned} \tag{3.1}$$

where Ω is bounded with a smooth boundary, $\beta \geq 0$ is a C^2-function, and the function f satisfies the growth conditions (2.2) and the dissipative condition (2.3) with λ_1 being the first eigenvalue of $-\Delta$ with homogeneous Dirichlet conditions.

Our presentation follows Hale & Raugel (1993a), which is based heavily on Hale (1988). In this space, we cannot give too many details and can only suggest that the reader consult these references and the others mentioned below.

Let us write the equation as a system

$$\begin{aligned} &U_t = CU + \tilde{f}(U) \\ &C = \begin{bmatrix} 0 & I \\ -\Delta & -\beta \end{bmatrix} \qquad \tilde{f}(U) = \begin{bmatrix} 0 \\ f(u) \end{bmatrix} \qquad U = \begin{bmatrix} u \\ u_t \end{bmatrix} \end{aligned} \tag{3.2}$$

If we choose initial data $U_0 = (\varphi, \psi)$ for (3.2) in the space $H^1(\Omega)\times L^2(\Omega)$, then any solution is given by the variation of constants formula

$$U(t) = e^{Ct}U_0 + \int_0^t e^{C(t-s)}\tilde{f}(U(s))ds, \quad t \geq 0. \tag{3.3}$$

Local existence of solutions can be obtain by using this formula. To obtain global existence, we use the energy functional

$$V(\varphi,\psi) = \int_\Omega [\frac{1}{2}|\nabla\varphi|^2 + \frac{1}{2}|\psi|^2 - F(\varphi)]dx, \tag{3.4}$$

where $F(u) = \int_0^u f(s)ds$. As for the parabolic equation in the previous section, one shows that

$$\dot{V}(u(t), u_t(t)) = -\int_\Omega \beta(x)|u_t(x,t)|^2 \leq 0. \tag{3.5}$$

Using the fact that $V(\varphi,\psi) \to \infty$ as $|(\varphi,\psi)| \to \infty$, we see that solutions exists globally and, in fact, $\gamma^+(B)$ is bounded for any bounded set $B \subset H^1(\Omega) \times L^2(\Omega)$.

If we let $T(t)(\varphi,\psi)$ denote the solution of (3.1) through (φ,ψ), then $T(t)$ is C^1-group on $H^1(\Omega) \times L^2(\Omega)$.

Let us next prove the following result.

Theorem 3.1. *If we suppose that there are positive constants k, α such that*

$$|e^{Ct}|_{\mathcal{L}(H^1(\Omega)\times L^2(\Omega), H^1(\Omega)\times L^2(\Omega))} \leq ke^{-\alpha t}, \quad t \geq 0, \tag{3.6}$$

and that $\gamma < 1$ for $n = 3$, then there is a locally compact minimal global attractor for (3.1). If, in addition, (3.1) is point dissipative, then there is a global attractor.

Under the assumption (2.2) and $\gamma < 1$ for $n = 3$, the mapping $f : H^1(\Omega) \to L^2(\Omega)$ is compact. This fact, together with (3.3) and the assumption (3.6) imply that the semigroup $T(t)$ is asymptotically smooth and the the first statement of the theorem follows from Lemma 1.3. The last statement follows from Theorem 1.2.

To apply Theorem 3.1, we must first obtain conditions which will imply that the linear system $\dot{U} = CU$ satisfies (3.6). If $\beta > 0$ is a constant, it is not difficult to show that this is the case. Furthermore, if $\beta > 0$ is a constant, then (3.5) implies that the ω-limit set of any positive orbit must be in the set of equilibrium points E, which is bounded by the dissipative condition (2.3). Therefore, there is a global attractor for (3.1) if $\beta > 0$ is a constant. This result proved in this way was given by Hale (1984). Haraux (1984) also obtained the same result with a different proof.

What happens when there is only local damping $\beta(x)$? More specifically, what conditions must be imposed on β in order to ensure that (3.6) is satisfied? If $\beta(x) > 0$ at some point $x_0 \in \Omega$, it was proved by Iwasaki (1969), with a simpler proof given by Dafermos (1978), that every solution of the linear equation $\dot{U} = CU$ approaches zero as $t \to \infty$. For $n = 1$, this is enough to show that the approach to zero is uniform with respect to initial data in a ball and therefore (3.6) is satisfied (see, for example, Hale & Raugel, 1993a). On the other hand, as remarked by Dafermos (1978), for $n \geq 2$, a result of Ralston (1969) can be used to give an example where the approach to zero is not uniform on a ball and so (3.6) is not satisfied.

The following result of Bardos *et al.* (1991) solved this problem in a nice geometric way as stated in the following result.

Theorem 3.2. *If there is a $\tau > 0$ such that every ray of geometric optics intersects the set $\beta_s \times (0, \tau)$, where β_s is the support of β, then (3.6) is satisfied. In particular, this is true if $\beta_s \supset \partial\Omega$.*

Theorem 3.2 gives a very interesting solution to the verification of (3.6). However, there is still the question of characterizing for a particular domain the minimal conditions on the damping β for which the hypotheses are satisfied. Some specific results have been obtained by Chen *et al.* (1991).

In order to obtain a global attractor, we must also be concerned with the verification of point dissipativeness. From (3.5), we deduce that $\omega(\varphi, \psi) \subset G$, where G is the set of globally defined solutions of the problem

$$\begin{aligned} v_{tt} - \Delta v &= f(v) \quad \text{in } \Omega \setminus \beta_s, \\ v_t &= 0 \quad \text{in } \beta_s, \\ v &= 0 \quad \text{in } \partial\beta_s. \end{aligned} \tag{3.7}$$

If G consisted only of equilibrium points, then we would have point dissipativeness and thus the existence of a global attractor. If $n = 1$, it is not too difficult to show that this is the case (Hale & Raugel, 1993a).

If $n > 1$, by using L^n regularity and setting $w = v_t$, Hale & Raugel (1993a) show that the set G consists only of equilibrium points if, for each $\tau > 0$, $w = 0$ is the only weak $L^2(\Omega \times (0, \tau))$ solution of the following problem:

$$\begin{aligned} w_{tt} - \Delta w + b(x,t)w &= 0 \quad \text{in } \Omega \times (0, \tau), \\ w &= 0 \quad \text{in } \beta_s \times (0, \tau), \\ b &\in L^\infty((0, \tau); L^n(\Omega)). \end{aligned} \tag{3.8}$$

Knowing that there exists a $\tau > 0$ such that the only solution of (3.8) is $w = 0$ is referred to as the unique continuation property (u.c.p.) by Lions (1988) and was encountered in control theory. It is interesting to note that we arrive at a version of the u.c.p. from the consideration of dynamical properties of a flow rather than control theory. The u.c.p. will imply what we desire, but it is not known if our requirement for all $\tau > 0$ is equivalent to the u.c.p.

Ruiz (1992) has shown that the u.c.p. holds if $\beta_s \supset \partial\Omega$. As a consequence of this fact and Theorem 3.2, we obtain the existence of a global attractor if $\beta_s \supset \partial\Omega$.

Let us now turn to a discussion of the critical growth exponents on f, $\gamma = 2$ for $n = 2$, $\gamma = 1$ for $n = 3$. Nothing seems to be known about the case $\gamma = 2$, $n = 2$. For $\gamma = 1$, $n = 3$, the mapping $f : H^1(\Omega) \to L^2(\Omega)$ is a bounded map but is not compact. For this case and the additional assumption that $\beta > 0$ is a constant, Arrieta *et al.* (1992) gave the first proof of the existence of a global attractor by using a special decomposition of the nonlinearity f, a nonlinear variation of constants formula and some special properties of linear hyperbolic problems. Feireisl & Zuazua (1993) extended the result for $\gamma = 1$, $n = 3$, to the situation of local damping β under the hypothesis that $\beta_s \supset \partial\Omega$.

Recently, the existence of a global attractor has been proved for some special boundary conditions and special domains for the case where $\beta > 0$ is a constant, $n = 3$ and the exponent $\gamma \in [0, 3)$; that is, the growth rate on f can be almost 5. Lopes (1988 & 1990) did this for periodic boundary conditions by extending the problem to all of $\mathbb{R}^3$ and then using L^p, L^q regularity results. Kapitanski (1994) has done this for Ω a smooth manifold without boundary by employing the Strichartz inequalities.

4. Dependence on parameters

In this section, we are going to consider a family of semigroups $T_\epsilon(t)$ defined on a Banach space X_ϵ, where ϵ is a real parameter in $[0, \epsilon_0]$. We also assume that, for each ϵ in the half open interval $(0, \epsilon_0]$, there is a global attractor $\mathcal{A}_\epsilon$ for the semigroup $T_\epsilon(t)$. The basic question that we want is address is the following: How much information can we obtain about the flow defined by $T_0(t)$ from knowledge only of the flow defined by $T_\epsilon(t)$ and the manner in which $T_\epsilon(t)$ converges to $T_0(t)$ as $\epsilon \to 0$?

To be more specific, let us first discuss the case where $X_\epsilon = X_0$ for all ϵ and the hypothesis (H):

$$T_\epsilon(t)x - T_0(t)x \to 0 \text{ as } \epsilon \to 0 \text{ uniformly on bounded sets of } [0, \infty) \times X$$

holds.

Let us first assume that $T_0(t)$ also has a global attractor $\mathcal{A}_0$. If the flow on $\mathcal{A}_0$ is structurally stable and hypothesis (H) is satisfied with the convergence being in the C^1-topology, then we know that, for any fixed t, the flow defined by the map $T_\epsilon(t)$ on $\mathcal{A}_\epsilon$ is topologically equivalent to the flow defined by $T_0(t)$ on $\mathcal{A}_0$ (Hale *et al.*, 1984).

If we do not assume that we know so much about the flow on $\mathcal{A}_0$, then there are natural questions to ask:

1. Are the attractors $\{\mathcal{A}_\epsilon, \epsilon \in [0, \epsilon_0]\}$ upper semicontinuous at $\epsilon = 0$?
2. Are the attractors $\{\mathcal{A}_\epsilon, \epsilon \in [0, \epsilon_0]\}$ lower semicontinuous at $\epsilon = 0$?

We recall that $\{\mathcal{A}\epsilon\}$ is upper semicontinuous at $\epsilon = 0$ if

$$\lim_{\epsilon \to 0} \mathrm{dist}\,(\mathcal{A}_\epsilon, \mathcal{A}_0) = 0,$$

where $\mathrm{dist}\,(A, B) = \sup_{x \in A} \inf_{y \in B} |x - y|$. The family of sets $\{\mathcal{A}\epsilon\}$ is lower semicontinuous at $\epsilon = 0$ if $\lim_{\epsilon \to 0} \mathrm{dist}\,(\mathcal{A}_0, \mathcal{A}_\epsilon) = 0$.

Upper semicontinuity is easy to prove if (H) is satisfied and is an almost immediate consequence of the strong stability property of $\mathcal{A}_0$, the semigroup property and the invariance of the attractors. We obtain upper semicontinuity without assuming anything about the flow on the attractor $\mathcal{A}_0$.

Lower semicontinuity of the attractors requires some type of hypothesis about the flow on $\mathcal{A}_0$. The simple example $\dot{y} = (1 - y)(\epsilon + y^2)$ illustrates this fact.

Some general results on lower semicontinuity are known for gradient systems, special cases of which we have discussed in the previous sections. For example, if (H) is satisfied and the equilibrium points of $T_0(t)$ are hyperbolic, then the attractors are lower semicontinuous at $\epsilon = 0$. With a stronger hypotheses on the convergence of $T_\epsilon(t)$ to $T_0(t)$, one can obtain estimates on the rate of convergence of the attractors (Hale & Raugel, 1989).

In many applications, the spaces X_ϵ are not all the same and then we have to first determine how to relate the spaces in order to have a concept of convergence of the semigroups which replaces the hypothesis (H). Also, the dependence upon the parameter may not be so nice. We gave one illustration of this in Section 2 in our discussion of small viscosity in a nonhomogeneous conservation law. In that case, we were able to prove upper semicontinuity and almost proved continuity of the attractors by considering all problems in the same space L^p even though the basic flow was in other spaces which were completely different for $\epsilon = 0$ and $\epsilon > 0$.

As another illustration, let us consider the singularly perturbed wave equation

$$\epsilon^2 u_{tt} + u_t - \Delta u = f(u) \text{ in } \Omega \tag{4.1}$$

with homogeneous Dirichlet boundary conditions, where Ω is a smooth bounded domain and f satisfies growth conditions (2.2) and dissipative condition (2.3) with λ_1 being the first eigenvalue of $-\Delta$. Let $\mathcal{A}_\epsilon$ be the global attractor of (4.1) in the space $H^1(\Omega) \times L^2(\Omega)$.

For $\epsilon = 0$, the parabolic equation defines a semigroup $\tilde{T}(t)$ on the space $H^1(\Omega)$ with a global attractor $\tilde{\mathcal{A}}_0$. Using the regularity properties of $\tilde{\mathcal{A}}_0$, namely, that $\tilde{\mathcal{A}}_0 \subset H_0^1(\Omega) \times H^2(\Omega)$, we can define the set $\mathcal{A}_0 \subset H^1(\Omega) \times L^2(\Omega)$ by the relation

$$\mathcal{A}_0 = \{(\varphi, \psi) : \varphi \in \tilde{\mathcal{A}}_0, \psi = \Delta\varphi - f(\varphi)\}.$$

Through this artifice, we now have the attractors $\mathcal{A}_\epsilon, \epsilon \in [0, \epsilon_0]$ in the same space. Hale & Raugel (1988) then show that the these attractors are upper semicontinuous in ϵ at $\epsilon = 0$. Furthermore, if the equilibrium points for $\epsilon = 0$ are hyperbolic, then they are lower semicontinuous (Hale & Raugel (1990)) and therefore Hausdorff continuous at $\epsilon = 0$.

The proof of the upper semicontinuity involves obtaining first the a priori estimates for all $\epsilon \in (0, \epsilon_0]$:

$$\mathcal{A}_\epsilon \subset B,$$

where B is a bounded set in $(H^2(\Omega) \cap H_0^1(\Omega)) \times H^1(\Omega)$ and there is a constant C such that

$$\epsilon |u_{tt}^\epsilon|_{L^2} \leq C$$

for all solutions $(u^\epsilon, u_t^\epsilon) \in \mathcal{A}_\epsilon$, $t \in \mathbb{R}$. With these estimates, we have a regular perturbation of the parabolic equation from which we obtain the upper semicontinuity (see Hale & Raugel (1988) for the details). For the lower semicontinuity, it is necessary to obtain estimates relating the unstable manifolds of the equilibrium points for $\epsilon = 0$ and $\epsilon \neq 0$. Also, it is necessary to obtain estimates on finite time intervals between the solution u^ϵ of the hyperbolic equation and the solution u^0 of the parabolic equations (see Hale & Raugel (1990) for the details).

Another instance in which the dependence upon the parameter is not nice is the case of thin domains. We illustrate this only for the first nonlinear problem of thin domains that was discussed, namely, the parabolic equation (2.1) following Hale & Raugel (1992b). The ideas which were begun in that paper have been further developed and modified to apply to many other situations; for example, damped hyperbolic equations, the Navier-Stokes equation on thin three dimensional domains and the Euler equation on a thin region between two spheres. We do not discuss these problems and refer the reader to the extensive C.I.M.E. lecture notes of Raugel (1995).

To make the presentation as simple as possible, we consider a very special case of the results in Hale & Raugel (1992b). Let

$$Q_\epsilon = \{(X, Y) \in \mathbb{R}^2 : 0 < Y < \epsilon g(X), X \in (0, 1)\},$$

where $\epsilon > 0$ is a parameter and g is a C^3-function on $[0, 1]$. We call Q_ϵ a thin domain over the interval $(0, 1)$ and consider the parabolic equation

(2.1) on Q_ϵ. The objective is to determine an equation on $(0,1)$ which is a natural limit of (2.1) as $\epsilon \to 0$ and for which the dynamics of the limiting equation will coincide with the dynamics of (2.1) on Q_ϵ, at least if the flow on the limit equation is structurally stable.

To discuss this problem, it is natural to scale Q_ϵ to the square $Q = (0,1) \times (0,1)$ by the transformation $X = x$, $Y = \epsilon g(x) y$ to obtain a new partial differential equation. By using the variational characterization of the new PDE, Hale & Raugel (1992b) show that the natural limit equation is

$$u_t = \frac{1}{g}(g u_x)_x + f(u) \tag{4.2}$$

with $u_x = 0$ at $x = 0, 1$.

The scaled equation (2.1) has a global attractor $\mathcal{A}_\epsilon$ in the space $H^1_\epsilon(Q)$ which is $H^1(Q)$ with the norm

$$\|\varphi\|^2_{H^1_\epsilon(Q)} = \|\varphi\|^2_{H^1(Q)} + \|\varphi\|^2_{L^2(Q)}.$$

The equation (4.2) has a global attractor $\tilde{\mathcal{A}}_0$ in $H^1((0,1))$ which has a natural embedding into $\mathcal{A}_0$ in $H^1(Q)$ by defining $\psi \in \mathcal{A}_0$ as $\psi(x,y) = \varphi(x)$ for $\varphi \in \tilde{\mathcal{A}}_0$.

Hale & Raugel (1992b) prove that the attractors $\mathcal{A}_\epsilon, \epsilon \in [0, \epsilon_0]$ are upper semicontinuous in $\epsilon = 0$ in $H^1_\epsilon(Q)$. There are two main ideas in the proof. The first is to introduce the mean value operator $M : H^1(Q) \to H^1((0,1))$ by the relation

$$(M\varphi)(x) = \int_0^1 \varphi(x,y)dx.$$

The second is to obtain good estimates on finite time intervals for the expression $T_\epsilon(t)\varphi - T_0(t)M\varphi$ for $\varphi \in H^1(Q)$, where $T_\epsilon(t)$ is the semigroup defined by (2.1) and $T_0(t)$ is the semigroup defined by (4.2).

Hale & Raugel (1992b) also show that, if the limit equation is structurally stable, then the flow of (2.1) and (4.2) are topologically equivalent on the attractors. This is proved by showing that there is an inertial manifold for (2.1) which is described by ODE for which the vector field is continuous in ϵ in the C^1-topology of vector fields. Detailed spectral analysis is required to prove this together with the theory of invariant manifolds.

There are many other situations for which the dependence on the parameters is not nice. For example, this is the case for either discretization of the PDE in space or time. However, we will not further enlarge on this topic but refer to the above references as well as the work of Stuart & Humphries (1996) and the references therein.

Up to this point, we have assumed that the semigroup $T_0(t)$ is dissipative and possesses a global attractor. What if $T_0(t)$ is conservative? How much information can we obtain about a conservative system by considering the limit of dissipative systems by allowing the dissipation to go to zero? This is certainly an important problem and is the subject of many investigations. We cannot hope to obtain too many specific properties of the dynamics of the limit system, but there should be a possibility of obtaining some type of general information about the manner in which the orbits of the dissipative systems wander over the level sets of the energy of the limit system.

Hale & Raugel (1993b) have introduced a concept which at least is a first attempt at studying the above problem. Again, to make the presentation simple, we assume that all of the Banach spaces are the same, $X_\epsilon = X$.

Definition 4.1. For any set $B \subset X$, $\epsilon_0 > 0$, we say that $y \in \hat{\omega}_{(0,\epsilon_0]}(B)$ if there are sequences $\epsilon_n \to 0$, $t_n \to \infty$ as $n \to \infty$ and $\{y_n\} \subset B$ such that $T_{\epsilon_n}(t_n)y_n \to y$ as $n \to \infty$.

Notice that the definition of $\hat{\omega}_{(0,\epsilon_0]}(B)$ treats ϵ as if it were also a time parameter. A trivial example, but one that also captures the spirit of the definition is the scalar equation $\dot{x} = -\epsilon x$, $\epsilon \geq 0$. For each $\epsilon > 0$, the global attractor $\mathcal{A}_\epsilon$ is zero. For the set $B = [0,1]$ we have $\hat{\omega}_{(0,\epsilon_0]}(B) = B$. For $\epsilon = 0$, the equation $\dot{x} = 0$ has a first integral $V(x) = x$ and the ω-limit set $\omega_0(B)$ for this equation is B. Therefore, $\hat{\omega}_{(0,\epsilon_0]}(B) = \omega_0(B)$. Therefore, by taking this special type of limit, we recover from the dissipative system all of the limits of the conservative system.

The above limit also gives information in the dissipative case without assuming anything about the attractor for the limit system. For example, for the equation $\dot{x} = (\epsilon + x^2)(1 - x)$, for $\epsilon > 0$, $\mathcal{A}_\epsilon = \{0\}$ and for $\epsilon = 0$, $\mathcal{A}_0 = [0,1]$. We obtain no information about the attractor at $\epsilon = 0$ by taking the limit of the attractors for $\epsilon > 0$. On the other hand, if we take $B = [-1,2]$, then $\hat{\omega}_{(0,\epsilon_0]}(B) = \mathcal{A}_0$. We recover the attractor for the limit equation without making additional hypotheses on the flow.

These examples are typical of the types of general results proved in Hale & Raugel (1993b). We state two particular results from that paper.

Theorem 4.1. *Suppose that hypothesis (H) is satisfied and each $T_\epsilon(t)$, $\epsilon \in [0,\epsilon_0]$, has a global attractor $\mathcal{A}_\epsilon$. If B is a bounded set which contains $\mathcal{A}_0$, then $\hat{\omega}_{(0,\epsilon_0]}(B) = \mathcal{A}_0$.*

This result is applicable, for example, to the parabolic equation in Section 2 and the linearly damped wave equation in Section 3 where we allow the nonlinear function f to be of the form $f = f(u,\epsilon)$ with f being smooth in u, ϵ. Without making any assumptions about the hyperbolicity of the equilibrium points we obtain $\hat{\omega}_{(0,\epsilon_0]}(B) = \mathcal{A}_0$ provided that $\mathcal{A}_0 \subset B$. The set B can be chosen to the set where the energy is $\leq c$, where c is a sufficiently large constant.

The following result applies when the system at $\epsilon = 0$ is conservative.

Theorem 4.2. *Suppose that (H) is satisfied, $T_0(t)$ is a group, and is conservative with first integral $V(x)$ which approaches ∞ as $|x| \to \infty$ and $V(T_\epsilon(t)x) \le V(x)$ for all $t \ge 0$, $\epsilon \in [0, \epsilon_0]$ and $x \in X$. If $B_c = \{x : V(x) \le c\}$ and $\cup_{\epsilon \in (o,\epsilon_0]} \mathcal{A}_\epsilon \subset B_c$, then $\hat{\omega}_{(0,\epsilon_0]}(B_c) = \mathcal{A}_0(B_c)$, where $\mathcal{A}_0(B_c)$ is the ω-limit set of $B - c$ under the semigroup $T_0(t)$.*

As an example, take $\epsilon = \beta$ and consider the system (4.1) with $\beta \ge 0$ a constant. The energy function V in (3.4) does not depend upon β and is non-increasing along the solutions. For $\beta > 0$, we have seen that there is a global attractor $\mathcal{A}_\beta$ and these attractors are uniformly bounded. Therefore, we can find a $c > 0$ such that the conditions of Theorem 4.2 are satisfied. The equation $\beta = 0$ is conservative with the energy function V in (3.4). Therefore, the conclusion of Theorem 4.2 holds. In this case, $\mathcal{A}_0(B_c) = B_c$.

As another simple application, consider the ODE $\dot{u} = v$, $\dot{v} = -f(u) - \beta v$, where $\beta \ge 0$. The energy is $V(u, v) = \frac{1}{2}v^2 + F(u)$, $F(u) = \int_0^u f(s)ds$. The above theorem applies directly to this situation to obtain $\hat{\omega}_{(0,\epsilon_0]}(B_c) = B_c$ if c is sufficiently large. It also is possible to discuss the limits of the attractors as $\beta \to 0$ (see Hale & Raugel, 1993b). For the special case, $f(u) = -u + u^3$, the limits of the attractors is the closure of the interior of the figure eight in the (u, v)-plane consisting of homoclinic orbits to the origin.

Another application discussed in Hale & Raugel (1993b) is the equation

$$\dot{x} = -(1 + \epsilon)f(x(t)) + f(x(t - 1))$$

where $\epsilon \ge 0$, $f(0) = 0$, $f'(x) \ge \delta > 0$ for all x. This equation is considered in the space $X = C([-1, 0], \mathbb{R})$. For any initial data in $\varphi \in X$, there is a unique solution $x(t, \varphi)$ defined for $t \ge -1$. The semigroup $T_\epsilon(t)$ is defined by the relation $(T_\epsilon(t)\varphi)(\theta) = x(t + \theta, \varphi)$, $\theta \in [-1, 0]$. It is shown that $\mathcal{A}_\epsilon = \{0\}$ for each $\epsilon > 0$. On the other hand, for $\epsilon = 0$, the function $V(\varphi) = \varphi(0) + \int_{-1}^0 f(\varphi(s))ds$ is a first integral. On each set $V^{-1}(c)$, there is a unique equilibrium point e_c and $\mathcal{A}_0(V^{-1}(c)) = \{e_c\}$. It is then shown that $\hat{\omega}_{(0,\epsilon_0]}(B) = \omega_0(B) = I(B)$, where $I(B) = \{e(c_\varphi) : \varphi \in B\}$, where $V(\varphi) = c_\varphi$.

Another interesting example discussed in Hale & Raugel (1993b) is the ODE

$$\dot{\xi} = f(\xi) - \eta, \quad \dot{\eta} = \epsilon(\gamma\xi - \eta),$$

where $\epsilon \ge 0$, $\gamma > 0$, $f(u = u(1 - u)(u - a)$, $0 < a < 1/2$. The constant γ is chosen so that there are three equilibrium points, two stable hyperbolic and one hyperbolic saddle. For $\epsilon > 0$, the system is dissipative and the attractor $\mathcal{A}_\epsilon$ consists of the unstable manifold of the saddle and the two stable equilibrium points. The limit of the attractors is a horizontal segment with two small segments on the cubic curve $C := \{(\xi,\eta) : \eta = f(\xi)\}$. For $\epsilon = 0$, the system is conservative with first integral $V(\xi,\eta) = \eta$. For fixed η, $\omega_0(V^{-1}(\eta))$ is either a point or a line segment. Let $\tilde{C} = \cup\,\omega_0(V^{-1}(\eta))$ over line segments. If we consider a set B which is the closure of an open set which contains $\tilde{C}$, then $\hat{\omega}_{(0,\epsilon_0]}(B) = \omega_0(B)$.

The equation

$$\begin{aligned} u_t &= \delta u_{xx} + f(u) - v \\ v_t &= \epsilon(\gamma u - v) \end{aligned}$$

also is considered in Hale & Raugel (1993b). This system is dissipative if $\epsilon > 0$ and is conservative if $\epsilon = 0$. They show that $\hat{\omega}_{(0,\epsilon_0]}(B) = \omega_0(B)$ if B is a sufficiently large set. If δ is sufficiently large, then this limit is the same as thc onc for the ODE above.

As a final remark, we point out that it is shown that, for dissipative equations as considered in Sections 2 and 3 on thin domains, Hale and Raugel have shown that $\hat{\omega}_{(0,\epsilon_0]}(B) = \mathcal{A}_0$ if B contains $\mathcal{A}_0$ without any assumption about the flow on $\mathcal{A}_0$.

5. Acknowledgement

These lecture notes were completed with partial support from NSF and ARO and while the author was a visiting Rothschild Professor at the Newton Institute for Mathematical Sciences, Cambridge.

References

Alikakos, N.D., Bates, P. & Fusco, G. (1991) Slow motion for the Cahn-Hilliard equation in one space dimension. *J. Differential Equations* **90**, pp. 81–135.

Alikakos, N.D. & Fusco, G. (1992) Equilibrium and dynamics of bubbles for the Cahn-Hilliard equation, Preprint.

Angenent, S. (1986) The Morse-Smale property for a semi-linear parabolic equation. *J. Differential Equations* **62**, pp. 427–442.

Arrieta, J., Carvalho, A.N. & Hale, J.K. (1992) A damped hyperbolic equation with a critical exponent, *Comm. PDE* **17**, pp. 841–866.

Babin, A.V. & Vishik, M.I. (1991) *Attractors of Evolution Equations*. North Holland.

Bardos, C., Lebeau, G. & Rauch, J. (1992) Sharp sufficient conditions for the observation, control and stabilization of waves from the boundary, *SIAM J. Control Optim.* **30**, pp. 1024–1065.

Bates, P.W. & Xun, J. (1994) Metastable patterns for the Cahn-Hilliard equation. Part 1, *J. Differential Equations* **111**, pp. 421–457.

Bates, P.W. & Xun, J. (1995) Metastable patterns for the Cahn-Hilliard equation. Part 2, *J. Differential Equations* **117**, pp. 165–216.

Bochner, S. (1962) A new approach to almost periodicity, *Proc. Nat. Acad. Sci.* **48**.

Brunovský, P. & Poláčik, P. (1997) Morse-Smale structure of generic reaction diffusion equations in higher space dimension, *J. Differential Equations* **135**, pp. 129–181.

Brunovský, P., Poláčik, P. & Sanstede, B. (1992) Convergence in general periodic parabolic equations in one space dimension, *Nonlinear Analysis* **18**, pp. 209–215.

Carr, J. & Pego, R.L. (1989) Metastable patterns in solutions of $u_t = \epsilon^2 u_{xx} - f(u)$ for small ϵ. *Comm. Pure Appl. Math.* **42**, pp. 523–576.

Carr, J. & Pego, R.L. (1990) Invariant manifolds and metastable patterns in solutions of $u_t = \epsilon^2 u_{xx} - f(u)$ for small ϵ, *Proc. Roy. Soc. Edinburgh* **116**, pp. 133–160.

Casten, R.C. & Holland, C.J. (1978) Instability results for reaction diffusion equations with Neumann boundary conditions, *J. Differential Equations* **27**, pp. 266–273.

Chafee, N. & Infante, E.F. (1974) A bifurcation problem for a nonlinear parabolic equation, *Applicable Analysis* **4**, pp. 17–37.

Chen, G., Fulling, S.A., Narcowich, F.J. & Sun, S. (1991) Exponential decay of energy of evolution equations with locally distributed damping, *SIAM J. Appl. Math.* **51**, pp. 266–301.

Chen, W., Chen, X.-Y. & Hale, J.K. (1992) Structural stability for time-periodic one-dimensional parabolic systems, *J. Differential Equations* **96**, pp. 355–418.

Chen, X.-Y. & Matano, H. (1989) Convergence, asymptotic periodicity, and finite-point blow-up in one-dimensional semilinear heat equations, *J. Differential Equations* **78**, pp. 160–190.

Chen, X.-Y. & Poláčik, P. (1995) Gradient-like structure and Morse decompositions for time-periodic one-dimensional parabolic equations, *J. Dyn. Diff. Eq.* **7**, pp. 73–107.

Dafermos, C. (1978) Asymptotic behavior of solutions of evolutionary equations. In *Nonlinear Evolution Equations, Ed. M.G. Crandall*, pp. 103–123, Academic Press.

Dancer, N. & Hess, P. (1994) Stable subharmonics in reaction diffusion equations *J. Differential Equations* **108**, pp. 190–200.

Deimling, K. (1985) *Nonlinear functional analysis*. Springer-Verlag.

Fan, H. & Hale, J.K. (1994) Large time behavior in inhomogeneous conservation laws, *Arch. Rat. Mech. Anal.* **125**, pp. 201–216.

Fan, H. & Hale, J.K. (1995) Attractors in inhomogeneous conservation laws and parabolic regularization, *Trans. Am. Math. Soc.* **347**, pp. 1239–1254.

Feireisl, E. & Zuazua (1993) Global attractors for semilinear wave equations with locally distributed nonlinear damping and critical exponent, *Comm. PDE* **18**, pp. 1539–1556.

Fiedler, B. & Mallet-Paret, J. (1989) A Poincaré-Bendixson theorem for scalar diffusion equations, *Arch. Rat. Mech. Anal.* **107**, pp. 325–345.

Fiedler, B. & Rocha, C. (1996) Heteroclinic orbits of semilinear parabolic equations, *J. Differential Equations* **125**, pp. 239–281.

Fusco, G. (1990) A geometric approach to the dynamics of $u_t = \epsilon^2 u_{xx} + f(u)$ for small ϵ. In *Lect. Notes Physics* **359**, pp. 53–73.

Fusco, G. & Hale, J.K. (1989) Slow-motion manifolds, dormant instability and singular perturbations, *Dyn. Diff. Equations* **1**, pp. 75–94.

Fusco, G., Hale, J.K. & Xun, J. (1995) Traveling waves as limits of solutions on bounded domains, *SIAM J. Math. Anal.* **26**, pp. 21–34.

Fusco, G. & Oliva, W.M. (1988) Jacobi matrices and transversality, *Proc. Roy. Soc. Edinburgh* **109A**, pp. 231–243.

Fusco, G. & Rocha, C. (1991) A permutation related to the dynamics of a scalar parabolic PDE, *J. Differential Equations* **91**, pp. 75–94.

Grant, C. (1995) Slow motion in one dimensional Cahn-Morral systems, *SIAM J. Math. Anal.* **26**, pp. 21–34.

Hale, J.K. (1984) Asymptotic behavior and dynamics in infinite dimensions. In *Research Notes Math.* **132**, pp. 1–42, Pitman.

Hale, J.K. (1988) *Asymptotic Behavior of Dissipative Systems*. Math. Surveys and Monographs, Vol. 25, Am. Math. Soc.

Hale, J.K., Magalhães, L. & Oliva, W.M. (1984) *An Introduction to Infinite Dimensional Systems*. Appl. Math. Sci. **47**, Springer.

Hale, J.K. & Raugel, G. (1988) Upper semicontinuity of the attractor for a singularly perturbed hyperbolic equation, *J. Differential Equations* **73**, pp. 197–214.

Hale, J.K. & Raugel, G. (1989) Lower semicontinuity of attractors of gradient systems and applications, *Ann. Mat. Pura Appl. (IV)* **CLIV**, pp. 281–326.

Hale, J.K. & Raugel, G. (1990) Lower semicontinuity of the attractor for singularly perturbed hyperbolic equation, *J. Dyn. Diff. Eqns.* **2**, pp. 19–67.

Hale, J.K. & Raugel, G. (1992a) Convergence in gradient like systems and applications, *ZAMP* **43**, pp. 63–124.

Hale, J.K. & Raugel, G. (1992b) Reaction-diffusion equations on thin domains, *J. Math. Pures Appl.* **71**, pp. 33–95.

Hale, J.K. & Raugel, G. (1993a) Attractors for dissipative evolutionary equations. In Perelloó, Simó, & Solá-Morales (eds), *Differential Equations*, World Scientific, pp. 3–22.

Hale, J.K. & Raugel, G. (1993b) Limits of semigroups depending on parameters, *Resenhas IME-USP* **1**, pp. 1–45.

Hale, J.K. & Verduyn-Lunel, S. (1993) *Introduction to Functional Differential Equations*. Appl. Math. Sci., Vol. 99, Springer.

Haraux, A. (1984) Two remarks on dissipative hyperbolic problems. In *Sem. Coll. Fr.* (Ed. J.L. Lions), Pitman.

Haraux, A. (1989) *Semilinear Hyperbolic Problems in Bounded Domains*. Math. Reports **3**, Part 1, Harwood Acad. Publishers.

Henry, D. (1981) *Geometric Theory of Semilinear Parabolic Equations*. Lect. Notes Math. **840**, Springer.

Henry, D. (1985) Some infinite dimensional Morse-Smale systems defined by parabolic differential equations, *J. Differential Equations* **59**, pp. 165–205.

Hess, P. (1987) Spatial homogeneity of stable solutions of some periodic problems with Neumann boundary conditions, *J. Differential Equations* **68**.

Iwasaki, N. (1969) Local decay of solutions for symmetric hyperbolic systems with dissipative and coercive boundary conditions in exterior domains. *Pub. Res. Inst. Math. Sci., Kyoto Univ.* **5**, pp. 193–218.

Kapitanski (1994) Damped wave equation on a smooth manifold, *Comm. PDE*, To appear.

Keller, J. & Sternberg, P. (1993) Preprint.

Kuramura, Ei & Mimura, M. (1992) Very slow dynamics for some reaction-diffusion equations of activator inhibitor type, *Japan J. Indust. Appl. Math.* **9**, pp. 35–77.

Ladyzenskaya, O.A. (1987) On the determination of the minimal global attractors for the Navier Stokes equation and other differential operators, *Russian Mathematical Surveys* **42**, pp. 27–73.

Lions, J.L. (1988) Exact controllability, stabilization and perturbations for distributed systems. *SIAM Review* **30**, pp. 1–68.

Lopes, O. (1988) Asymptotic behaviour of solutions of a nonlinear wave equation, *Bol. Soc. Mat. Brazil* **19**, pp. 141–150.

Lopes, O. (1990) Compact attractor for a nonlinear wave equation with critical exponent, *Proc. Roy. Soc. Edinburgh* **115A**, pp. 61–64.

Lyberopoulos, A.N. (1990) Asymptotic oscillations of solutions of scalar conservation laws with convexity under the action of a linear excitation, *Quart. Appl. Math.* **48**, pp. 755–765.

Lyberopoulos, A.N. (1994) Large time structure of solutions of scalar conservation laws with a nonlinear source field. *Proc. Roy. Soc. Edinburgh* **124A**, pp. 589–607.

Mallet-Paret, J. (1976) Negatively invariant sets of compact maps and an extension of a theorem of Cartwright, *J. Differential Equations* **22**, pp. 331–348.

Mañé, R. (1981) On the dimension of the compact invariant sets of certain nonlinear maps, *Lect. Notes Math.* **898**, pp. 230–242.

Matano, H. (1978) Convergence of solutions of one-dimensional semilinear parabolic equations, *J. Math. Kyoto Univ.* **18**, pp. 221–227.

Matano, H. (1979) Asymptotic behavior and stability of solutions of semilinear diffusion equations, *Publ. Res. Inst. Math. Sci., Kyoto Univ.* **15**, pp. 401–458.

Matano, H. (1982) Nonincrease of the lap number of a solution for a one-dimensional semilinear parabolic equation, *J. Fac. Sci. Univ. Tokyo, Section 1A, Math.* **29**, pp. 401–441.

Nickel, K. (1962) Gestaltaussagen über Lösungen parabolischer Differentialgleichungen, *J. Reine Angew. Math.* **211**, pp. 78–94.

Oliva, W.M., de Oliveira, J.C.F. & Solá-Morales, J. (1993) An infinite dimensional Morse-Smale map, To appear.

Pinto, J. (1995) Slow motion manifolds for reaction diffusion equations. *Ph.D. Thesis*, Georgia Tech.

Poláčik, P. (1995) High dimensional ω-limit sets and chaos in scalar parabolic equations, *J. Differential Equations* **119**, pp. 24–53.

Poláčik, P. (1994) Transversal and nontransversal intersection of stable and unstable manifolds of reaction diffusion equations on symmetric domains, *Differential and Integral Equations* **7**, pp. 1527–1545.

Poláčik, P. & Rybakowski, K. (1994) Imbedding vector fields into Dirichlet BVP's, *Ann. Scuola Norm. Sup. Pisa*, To appear.

Poláčik, P. & Rybakowski, K. (1996) Nonconvergent bounded trajectories in semilinear heat equations, *J. Differential Equations* **124**, pp. 472–494.

Poláčik, P. & Tereščák, I. (1992) Convergence to cycles as typical asymptotic behavior in smooth strongly monotone discrete-time dynamical systems, *Arch. Rat. Mech. Anal.* **16**, pp. 339–360.

Ralston, J.V. (1969) Solutions of the wave equation with localized energy, *Comm. Pure Appl. Math.* **22**, pp. 307–323.

Raugel, G. (1995) Dynamics of Partial Differential Equations on Thin Domains. In *Lect. Notes Math.* **1609**, pp. 208-315.

Rocha, C. (1994) Bifurcations in discretized reaction-diffusion equations. *Resenhas IME-USP* **1**, pp. 1–16.

Ruiz, A. (1992) Unique continuation for weak solutions of the wave equation plus a potential, *J. Math. Pures Appl.* **71**, pp. 455–467.

Shen, W. & Yi, Y. (1995a) On minimal sets of scalar parabolic equations with skew-product structure, *Trans. Am. Soc.* **347**, pp. 4413–4431.

Shen, W. & Yi, Y. (1995b) Almost automorphy and skew-product semiflow. Preprint.

Smith, H. (1995) *Monotone Dynamical Systems*. Am. Math. Soc.

Stuart, A.M. & Humphries, A.R. (1996) *Dynamical Systems and Numerical Analysis*. Cambridge University Press.

Takáč, P. (1991) A construction of stable subharmonic orbits in monotone time-periodic dynamical systems, Preprint.

Temam, R. (1988) *Infinite Dimensional Dynamical Systems in Mechanics and Physics*. Appl. Math. Sci., Vol. 68, Springer.

Zelenyak, T.I. (1968) Stabilization of solutions of boundary value problems for a second order parabolic equation with one space variable. *Differential Equations* **4**, pp. 17–22.

NONLINEAR DYNAMICS OF EXTENDED SYSTEMS

P. COLLET
Centre de Physique Théorique
Laboratoire CNRS UPR 14
Ecole Polytechnique
F-91128 Palaiseau Cedex
France
email: collet@orphee.polytechnique.fr

1. Introduction

In this set of lectures I will describe recent results on the dynamics of some nonlinear parabolic partial differential equations in unbounded domains. As in statistical mechanics, it is natural to look at infinite domains as limits of large ones, hoping that boundary conditions do not play a significant role in some relevant questions. Approximations of large domains by unbounded ones have been used tacitly in the Physics literature for a long time, see for example the use of Fourier transforms instead of Fourier series. As we will see below, in an unbounded domain one often meets solutions with structures of well defined sizes, and it is reasonable to expect that the infinite domain approximation will be relevant when the size of the structures is much smaller than the size of the domain.

These lectures are organized as follows. First I will discuss the problem of existence of the semiflow defined by the evolution equations. This raises new questions in particular since one wants to deal with a large enough phase space. In section III I will describe some recent results about (weak) attracting sets for the dynamics. Section IV will be devoted to the construction of amplitude equations near instability of continuous spectra. Finally, in Section V, I will discuss some large time asymptotic problems.

2. Global Existence of solutions

The first mathematical question that needs to be settled is of course the existence and regularity of the solutions of the evolution equations, or in

J. C. Robinson and P. A. Glendinning (eds.),
From Finite to Infinite Dimensional Dynamical Systems, 113–143.

other words to ensure that the equations define a semi-flow. These equations are usually nonlinear parabolic systems sometimes with constraints (for example the incompressibility condition of the Navier Stokes equation). There is of course a huge literature on this problem in bounded domains, however we want to attack this question in an unbounded domain. As always with dynamical systems, in order to speak properly of the time evolution one needs first to decide on a phase space describing the instantaneous state of the system. In the case of bounded domains one usually considers Sobolev spaces. One can also use such phase spaces in problems with unbounded domains, however this leads to a rather restricted set of questions (which may nevertheless have some interest). The reason is that in some vague sense functions which belong to L^p spaces ($1 \leq p < \infty$) tend to zero at infinity. In particular such phase spaces miss periodic solutions, waves etc., and a lot of other interesting solutions which have a non trivial behaviour at infinity. In order to capture these solutions we will have to consider large enough phase spaces essentially derived from L^∞ (we will require bounded solutions on physical grounds). A reasonable phase space to start with is the space $\mathcal{B} = C_b^{0,u}(\mathbf{R}^n)$ of uniformly continuous bounded functions (equipped with the sup norm). It is easy to verify for example that the heat semi-group is well behaved in this phase space: $\mathcal{B}$ is invariant and we have convergence to the initial data when the time decreases to zero.

In these larger phase spaces, the global existence and regularity problem has to be investigated for the each particular evolution equation. There are very few results available in this context and I will only describe two methods to obtain such results.

The first one is the maximum principle. One gets an a priori estimate by comparing the solution to a simple known one. However the method can be applied only to a restricted class of equations. I will only formulate the principle in one space dimension and give an application to the dissipative Burgers equation. For more general formulations and results see Protter & Weinberger (1967) or Friedman (1964).

Parabolic Maximum Principle. *Let $u_1(x,t)$ and $u_2(x,t)$ be bounded real functions on $(x_1,x_2)\times[0,T]$ ($T>0$ and x_1 and x_2 may be infinite). Let $a(x,t)$ and $b(x,t,z)$ be continuous real valued functions on $(x_1,x_2)\times[0,T]$ and $(x_1,x_2)\times[0,T]\times\mathbf{R}$ respectively. Assume that on $(x_1,x_2)\times[0,T]$*

$$\partial_t u_1 - \partial_x^2 u_1 - a(x,t)\partial_x u_1 - b(x,t,u_1) \geq \partial_t u_2 - \partial_x^2 u_2 - a(x,t)\partial_x u_2 - b(x,t,u_2)$$

and $u_1(x,0) \geq u_2(x,0)$ on (x_1,x_2). Moreover, if x_1 (respectively x_2) is finite, assume also $u_1(x_1,t) \geq u_2(x_1,t)$ on $[0,T]$ (respectively $u_1(x_2,t) \geq u_2(x_2,t)$ on $[0,T]$). Then on $(x_1,x_2)\times[0,T]$, we have

$$u_1(x,t) \geq u_2(x,t)\ .$$

In other words, under the above hypothesis, if $u_1 - u_2$ is non negative at the initial time and at the boundary, it is non negative inside the space time domain.

We now give an application to the global existence of the solutions of the dissipative Burgers equation. This is an equation for a real function $u(x,t)$ with $x \in \mathbf{R}$ and $t \in \mathbf{R}^+$. The evolution equation is given by

$$\partial_t u = \partial_x^2 u + u\partial_x u \,. \tag{DB}$$

The local time existence theorem of bounded solutions in $\mathcal{B} = C_b^{0,u}(\mathbf{R})$ is an easy exercise using the contraction mapping principle for the integral equation

$$u(x,t) = G_t * u(\cdot,0)(x) - 1/2 \int_0^t G'_{t-s} * u^2(\cdot,s)\, ds \,,$$

where G is the heat kernel.

Note however that using this method one can also derive, locally in time, regularity properties of the solution. In order to continue the solution in time one needs an *a priori* estimate on the solution. We assume that it is known that the solution exists and is bounded and continuous on a time interval $[0,T]$ (for example, for T small one can use the local time existence result mentioned above). Let M be a bound on the modulus of the initial condition. We first apply the maximum principle to the functions $u_1 = M$, $u_2 = u$, $b = 0$, $a(x,t) = u(x,t)$, and conclude that on $\mathbf{R} \times [0,T]$ we have $u \leq M$. Repeating the argument with $u_1 = u$ $u_2 = -M$ and the same functions a and b we finally conclude that on $\mathbf{R} \times [0,T]$, $|u| \leq M$. We can now enlarge the time interval of existence boundedness and regularity beyond T using the local time existence result starting at time T. Iterating this argument one easily derive the global existence (and also uniqueness, boundedness and regularity) of the solution.

There is another related equation which is given by

$$\partial_t \varphi = \partial_x^2 \varphi + (\partial_x \varphi)^2/2 \,. \tag{IDB}$$

The factor $1/2$ in the last term can be scaled away but we will keep it here for convenience. If φ satisfies (IDB), it is immediate to verify that $u = \partial_x \varphi$ satisfies Burgers equation. This equation can be solved using the Cole-Hopf transformation, namely the function

$$S = e^{\varphi/2}$$

satisfies the heat equation. Using the maximum principle one can therefore obtain upper and lower bounds on the function S and therefore on the function φ. However this method cannot be used to study (DB) because if u does not tend to zero fast enough at infinity, its primitive may be unbounded. Also an estimate on the function φ is usually not enough to obtain an estimate on the derivative.

Another way to prove the global existence and regularity of the solutions of (IDB) is to use that $u = \partial_x \varphi$ satisfies Burgers equation. Therefore, if we assume that the initial data is bounded and has a bounded uniform uniform derivative, by the previous result we obtain a uniform a priori estimate on the derivative. We can now apply the maximum principle with $a = \partial_x \varphi$, $b = 0$ and compare the solution to a constant. We obtain a uniform a priori estimate for the function φ and proceed as above to prove global existence, boundedness and regularity.

As illustrated above the maximum principle is a powerful tool to obtain a priori estimates, but unfortunately it only applies to a certain type of equation. Another important example of a nonlinear parabolic equation in an unbounded domains is the complex Ginzburg-Landau equation which is given by

$$\partial_t u = (1 + i\alpha)\Delta u + u - (1 + i\beta)u|u|^2 \qquad (CGL)$$

where α and β are real parameters, $u(x, t)$ is a complex valued function of $x \in \mathbf{R}^n$ and $t \in \mathbf{R}^+$. We have chosen here a convenient scaling for the various coefficients of the equation, other equivalent forms can be found in the literature.

The maximum principle does not apply to complex functions, and we will have to use another method to prove a global existence result for (CGL). We first formulate the available result (Collet, 1994a; see also Ginibre & Velo, 1995, for a generalisation).

Theorem II.1. *If $n = 1$ or 2, or if $n = 3$ and either $\alpha\beta > 0$ or one of the numbers $\sqrt{1 + \alpha^2}$ and $\sqrt{1 + \beta^2}$) is smaller than 2, then if $u_0 \in C_b^{0,u}(\mathbf{R}^n)$, the solution of (CGL) with initial data u_0 exists for all time. Moreover, there is a positive number $C = C(\alpha, \beta)$ and a (finite) time $T_0 = T_0(\alpha, \beta, u_0)$ such that for $t > T_0$*

$$\sup_{x \in \mathbf{R}^n} |u(x, t)| \leq C \, .$$

Similar uniform estimates are also true for the derivatives of u; we will come back to this question later and prove a stronger result.

The restriction on the coefficients in dimension 3 comes from the fact that the presently available proof relies on a Sobolev inequality which is saturated in dimension 2 and false in higher dimension. This may be related to the following phenomena. If one fixes the ratio $\alpha/\beta = -1$, and lets α diverge, by adequately rescaling the equation one obtains as a formal limit the nonlinear Schrödinger equation which is known to have singular solutions in dimension 2 or more.

We will not prove the above Theorem in its full generality but we will only illustrate the method in dimension one. We will also only treat the case $\alpha = \beta = 0$,

$$\partial_t u = \Delta u + u - u|u|^2,$$

the general case being similar with a little more algebra. For these special values of the parameters, the equation is usually called the Ginzburg-Landau equation (GL for short). In bounded domains, existence, uniqueness and regularity results rely on so called energy estimates (Liapunov functionals). However, in unbounded domains with solutions which are essentially non trivial at infinity, the corresponding Liapunov functionals diverge. The idea is then to use local energy estimates. As for the diffusive Burgers equation, local time existence, boundedness and regularity of the solution is an easy exercise left to the reader, and we will only concentrate on obtaining a priori bounds on the solution.

In order to localize the energy, we will need adequate cut-off functions. We will use the following convenient one (among many others)

$$\varphi(x) = \begin{cases} (1-x^2)^4 & \text{if } |x| < 1 \\ 0 & \text{if } |x| \geq 1. \end{cases}$$

This function is differentiable and has compact support in the interval $[-1, 1]$. We will also use the translates of this function, namely we define

$$\varphi_a(x) = \varphi(x-a) .$$

All the estimates will be independent of a, and this is why at the end we will obtain estimates valid on the whole real x line from local estimates on finite intervals. The first and typical result of the method is the following.

Lemma II.2. *There is a number $C > 0$ such that if u is any bounded differentiable solution of (GL) on $[0, T]$ with bounded derivatives, we have for any $a \subset \mathbf{R}$*

$$\frac{1}{2}\partial_t \int \varphi_a |u|^2 \leq -\frac{1}{2}\int \varphi_a |u'|^2 - \int \varphi_a |u|^2 - \frac{1}{2}\int \varphi_a |u|^4 + C .$$

Proof. One easily gets using integration by parts

$$\frac{1}{2}\partial_t \int \varphi |u|^2 = -\int \varphi |u'|^2 - \Re\left(\int \varphi' \bar{u} u'\right) + \int \varphi |u|^2 - \int \varphi |u|^4 .$$

We now observe that

$$\int \varphi' \overline{u} u' = \int_{-1}^{1} \varphi' \overline{u} u' = \int_{-1}^{1} \frac{\varphi'}{\sqrt{\varphi}} u' \sqrt{\varphi} \overline{u}$$

which implies by Schwarz's inequality

$$\left| \int_{-1}^{1} \varphi' \overline{u} u' \right| \leq \left(\int_{-1}^{1} \varphi |u'|^2 \right)^{1/2} \left(\int_{-1}^{1} \frac{\varphi'^2}{\varphi} |u|^2 \right)^{1/2}$$

$$\leq 8 \left(\int_{-1}^{1} \varphi |u'|^2 \right)^{1/2} \left(\int_{-1}^{1} \sqrt{\varphi} |u|^2 \right)^{1/2}$$

since on $[-1,1]$ it is easy to verify that $\varphi'^2/\varphi \leq 64\sqrt{\varphi}$.

By completing the square, we conclude that

$$-\frac{1}{2} \int \varphi |u'|^2 - \Re \left(\int \varphi' \overline{u} u' \right)$$

$$\leq -\frac{1}{2} \int \varphi |u'|^2 + 8 \left(\int_{-1}^{1} \varphi |u'|^2 \right)^{1/2} \left(\int_{-1}^{1} \sqrt{\varphi} |u|^2 \right)^{1/2} \leq 32 \int_{-1}^{1} \sqrt{\varphi} |u|^2 .$$

On the other hand, it is easy to verify that since $\varphi \leq 1$ and $34s - s^2/2 \leq 600$

$$32 \int_{-1}^{1} \sqrt{\varphi} |u|^2 + 2 \int_{-1}^{1} \varphi |u|^2 - 1/2 \int_{-1}^{1} \varphi |u|^4 \leq 1200 .$$

Since the GL equation is invariant by translation, the same estimate holds (in particular with the same constant C) with φ replaced by φ_a and this proves the Lemma.

As a corollary we now explain how to transform this inequality into a global *a priori* estimate.

Corollary II.3. *Let*

$$y(t) = \sup_{a \in \mathbf{R}} \int \varphi_a(x) |u(x,t)|^2 \, dx .$$

Then with the notations of the previous Lemma

1. *If* $\inf_{t \in [0,T]} y(t) \geq 2C$, *the function* $y(t)$ *is decreasing (on* $[0,T]$*).*
2. *If there is a number* $T_* \in [0,T[$ *such that* $y(T_*) \leq 2C$, *then for any* $t \in [T_*, T]$ *we have* $y(t) \leq 2C$.

Proof. For $a \in \mathbf{R}$, let

$$y_a(t) = \int \varphi_a(x)|u(x,t)|^2 \, dx \, .$$

From Lemma II.2 we have obviously

$$\frac{dy_a(t)}{dt} \leq -2y_a(t) + 2C \, .$$

Therefore as long as $y_a(t) > 2C$ the function $y_a(t)$ is decreasing, and if for some $t_a \in [0,T[$ we have $y_a(t_a) \leq 2C$ this will remain so until T. The assertions 1) and 2) follow now immediately from the definition of the function y.

In order to prove a uniform bound on the solution we will use a Sobolev inequality, and we therefore need to control the local L^2 norm of the solution. This is done by establishing an a *priori* inequality similar to Lemma II.2. We give the result below, leaving the proof (similar to the proof of Lemma II.2) as an exercise to the reader. We first define another cut-off function ψ by

$$\psi(x) = \begin{cases} (1-x^2)^6 & \text{if } |x| < 1 \\ 0 & \text{if } |x| \geq 1. \end{cases}$$

We also define the translates ψ_a as before.

Lemma II.4. *There are two constants $\Gamma > 1$ and D such that if u is any bounded differentiable solution of (GL) on $[0,T]$ with bounded derivatives, for any $a \subset \mathbf{R}$ the function*

$$z_a(t) = \int \psi_a(x)|u'(x,t)|^2 \, dx + \Gamma \int \varphi_a(x)|u(x,t)|^2 \, dx$$

satisfies the differential inequality

$$\frac{dz_a(t)}{dt} \leq -2z_a(t) + D \, .$$

We now give the proof of Theorem II.1. Note that this proof relies only on Lemma II.4, we only gave Lemma II.2 to explain the method in the simplest situation. We first assume that existence, boundedness and regularity have been proven up to a certain time τ (for example by a local time existence theorem). As in Corollary II.3, we deduce that the non-negative function $z(t) = \sup_a z_a(t)$ decreases as long as it is larger than D. If it happens to reach this value, it will stay below at any later time. We now observe that on the interval $[-1/2, 1/2]$ we have $\varphi > 1/16$ $\psi > 1/64$. Therefore for any $t \in [0, \tau]$ we have, using a standard Sobolev inequality,

$$\sup_{x\in[a-1/2,a+1/2]} |u(x,t)|^2 \leq \int_{a-1/2}^{a+1/2} |u'(x,t)|^2 \, dx + 2\int_{a-1/2}^{a+1/2} |u(x,t)|^2 \, dx \; .$$

Taking the supremum over a we conclude that for any $t \in]0, \tau]$

$$\|u(\cdot,t)\|_\infty^2 \leq 64z(t) \; .$$

As in the case of the Burgers equation, this a priori estimate allows us to extend the solution on a uniform time interval beyond τ ($u(\cdot,\tau) \in C_b^{0,u}$) using the local time existence Theorem. We can now repeat the previous argument and finally extend the existence and regularity result to any finite time.

It is interesting to observe at this point that all the above estimates are in fact local, therefore they will also apply to bounded domains provided we are at some distance from the boundary.

We now come to the regularity properties of the solution. The equation being parabolic, the evolution will in fact regularize the initial condition. One can prove the following result (Collet, 1994b; Takac *et al.*, 1996).

Theorem II.5. *Under the hypothesis of Theorem 1 there are two positive numbers $K = K(\alpha, \beta)$ and $\delta = \delta(\alpha, \beta)$ such that if $u_0 \in C_b^{0,u}(\mathbf{R}^n)$, there is a positive number $T^\omega = T^\omega(\alpha, \beta, u_0)$ such that for $t > T^\omega$ the function $u(x,t)$ is the restriction to the real x axis of a function analytic in the strip $|\Im z| \leq \delta$ and with modulus bounded in that strip by K.*

Note that in general the analytic continuation of $u(x,t)$ outside the real axis does not satisfy the CGL equation. In order to prove the result one first writes down on the real x axis the parabolic nonlinear system satisfied by the real and the imaginary parts of $u(x,t)$. It is for this system that one proves the analytic extension, i.e. the real and the imaginary parts of $u(x,t)$ each extend analytically to the strip. However outside the real axis these two extensions may not be the real and imaginary parts of an analytic function. We give a short sketch of the proof for the so called real Ginzburg Landau equation in dimension one where this complication does not occur. The proof in the general case is similar. The real Ginzburg Landau equation is given by

$$\partial_t u = u'' + u - u^3 \; ,$$

where u is a real function of x and t. The global existence and regularity result for this equation can be proved by the same method as for the CGL. As an exercise on the maximum principle, one can prove that the condition $\|u\|_\infty$ is invariant by the evolution. The proof of analyticity will in fact rely on a local time existence theorem in an adequate space of analytic functions. We first rewrite the equation in the form of an integral equation, namely for $t > t_0 > 0$ we have

$$u(x,t) = G_{t-t_0} * u(\cdot, t_0) + \int_0^{t-t_0} G_\tau * u(\cdot, t-\tau)^3 d\tau \ , \tag{II.1}$$

where

$$G_\tau(y) = \frac{1}{\sqrt{4\pi\tau}} e^{-y^2/4\tau} e^\tau \ .$$

The important remark is that the function $G_\tau(y)$ is entire in y and satisfies the bound

$$|G_\tau(y+i\xi)| = \frac{1}{\sqrt{4\pi\tau}} e^{-y^2/4\tau} e^\tau e^{\xi^2/4\tau} \leq \frac{1}{\sqrt{4\pi\tau}} e^2 e^{-y^2/4\tau}$$

if $0 < \tau \leq 1$ and $|\xi| \leq \sqrt{\tau}$.

We now start with a fixed t_0 assumed to be larger than the number T_0 of Theorem 1. We are going to consider for $t > t_0$ functions $v(z,t)$ which are analytic and bounded in the strip $|\Im z| \leq \sqrt{t-t_0}$. From the above remark, the function $G_{t-t_0} * u(\cdot, t_0)$ satisfies this requirement (it is even entire) and we have a bound

$$\sup_{0<t-t_0\leq 1} \ \sup_{|\Im z|\leq\sqrt{t-t_0}} |G_{t-t_0} * u(\cdot,t_0)(z)| \leq e^2 \|u(\cdot,t_0)\|_\infty \ .$$

The integral term in equation (II.1) is slightly more difficult to control. The function

$$G_\tau * v(\cdot, t-\tau)^3(z) = \int G_\tau(z-\xi) v(\xi, t-\tau)^3 d\xi \ ,$$

is entire in z for $t - t_0 > \tau > 0$, but for τ small we will not get in this way a good enough estimate.

The idea is to move the contour in the above integral using the assumed analyticity of $v(\xi, t-\tau)$ in the variable ξ. Indeed we have assumed that $v(\xi, t-\tau)$ is analytic and bounded in the strip $|\Im\xi| \leq \sqrt{t-\tau-t_0}$. If $z = y + i\eta$ and $|\eta| \leq \sqrt{t-t_0}$ (with for example $\eta > 0$, the estimate for $\eta < 0$ being similar), we move the integration contour to the line

$$\Im\xi = \frac{\sqrt{t-\tau-t_0}\ \eta}{\sqrt{t-\tau-t_0}+\sqrt{\tau}} < \sqrt{t-\tau-t_0} \ .$$

With this choice it is then easy to verify that for $0 < \tau < t - t_0$ we have $|\Im\xi| < \sqrt{t-\tau-t_0}$ and $|\eta - \Im\xi| < \sqrt{\tau}$. It now follows that for $t - t_0 > \tau > 0$

$$\sup_{|\Im z|\leq\sqrt{t-t_0-\tau}} \left| G_\tau * v(\cdot, t-\tau)^3(z) \right|$$

$$\leq \mathcal{O}(1) \sup_{0\leq\zeta\leq t-t_0-\tau} \sup_{|\Im z|\leq\sqrt{t-t_0-\tau-\zeta}} |v(z,t-\zeta)|^3 .$$

With this estimate one can use in the usual way a contraction mapping principle to complete the proof of a local time existence and uniqueness result in the space of analytic functions described above. The theorem follows at once from the uniqueness property.

The same results can be established for other equations using adequate variants of the methods presented above. We will only mention for later reference the one dimensional Swift Hohenberg equation

$$\partial_t u = \eta u - (1+\partial_x^2)^2 u - u^3 \tag{S.H.}$$

where u is a real function of $x \in \mathbf{R}$ and $t \in \mathbf{R}^+$, $\eta \in \mathbf{R}$ is a parameter.

3. Attracting and invariant sets of CGL (in dimension 1)

I will now discuss some recent results of Mielke & Schneider (1995) about attracting sets using some slightly different topologies than the original paper. We refer to this original paper for more details and other results. One difficulty about the definition of attractors in extended systems is that some initial conditions have an empty ω limit set. This problem can also be traced to the fact that we have a dynamics on a non compact topological space.

First of all we leave to the reader to prove the following (easy) consequences of Theorems II.1 and II.5. We will denote by S_t the semiflow generated by (CGL).

Corollary III.1. *There are three positive numbers M, ρ and T which depend only of the coefficients α and β such that if $\mathcal{B}$ denotes the set of functions analytic in the strip of width ρ around the real axis and with a modulus bounded by M in this strip, then*

1. *for any $u \in C_b^{0,u}$ we have $S_t(u) \in \mathcal{B}$ for t large enough.*
2. *For any $\tau \geq 0$ we have $S_{T+\tau}(\mathcal{B}) \subset \mathcal{B}$.*

We can now prove the following result about the global attracting set of (CGL).

Theorem III.2. *With the notations of Corollary III.1, the set*

$$\mathcal{A} = \bigcap_{n=0}^{\infty} S_{nT}(\mathcal{B})$$

is a (non empty), closed, translation invariant, S_t invariant, globally attracting set in $L^2(I, dx)$ for any interval I (we will denote below by d_I the *distance associated to this Hilbert space). In other words, we have for any $t > 0$ $S_t(\mathcal{A}) \subset \mathcal{A}$, and for any $u \in C_b^{0,u}$ and any interval I*

$$\lim_{t \to +\infty} d_I(S_t(u), \mathcal{A}) = \lim_{t \to +\infty} \inf_{v \in \mathcal{A}} d_I(S_t(u), v) = 0 .$$

Note that in the above argument no uniformity of convergence is claimed. We refer to Collet & Eckmann (1992) for some interesting examples concerning the heat equation. In some sense compactness was recovered at the price of going to a weaker topology. This topology is however local which is a natural requirement for extended systems (local observations).

Proof. First of all, $\mathcal{A}$ is not empty because it contains the zero solution. It is translation invariant because $\mathcal{B}$ is and since the evolution equation does not depend explicitly on space, the semiflow S_t commutes with translations. We leave to the reader to prove that for a fixed I and a fixed $n \geq 0$, the set $S_{nT}(\mathcal{B})$ is closed and compact in the topology of $L^2(I, dx)$ (use Montel's Theorem, see for example Hille, 1962), and therefore $\mathcal{A}$ is closed. We next observe, using 2) of Corollary III.1, that we have for any $t > 0$ and $n > 1$ $S_{t+nT}(\mathcal{B}) \subset S_{(n-1)T}(\mathcal{B})$. This implies immediately that $S_t(\mathcal{A}) \subset \mathcal{A}$ (since the sequence of sets $S_{nT}(\mathcal{B})$ is decreasing). We then observe that a function v belongs to $\mathcal{A}$ if and only if for any integer $n \geq 0$ there is a function $u_n \in \mathcal{B}$ such that $v = S_{nT}(u_n)$. Let $t > 0$ be given, and define a sequence (w_n) by

$$w_n = S_{2T+T[t/T]-t}\left(u_{n+2+[t/T]}\right)$$

where $[\cdot]$ denotes the integer part. We have for any fixed m and n large enough

$$v = S_t(S_{nT}(w_n)) = S_t(S_{mT}(S_{(n-m)T}(w_n))) .$$

Therefore for a fixed m, using compactness in each $L^2(I, dx)$ and the diagonal process, we can find a sequence of integers (n_j) such that for each interval $[-q, q]$ ($q \in \mathbf{N}$) the sequence

$$S_{(n_j-m)T}(w_{n_j})$$

converges in each $L^2(I, dx)$ to an element of $\mathcal{A}$. This proves that for any integer m

$$v = S_t(S_{mT}(v_m))$$

for some $v_m \in \mathcal{B}$ which implies that $\mathcal{A} \subset S_t(\mathcal{A})$ and we have proved the invariance of $\mathcal{A}$.

Finally, suppose there is a function $u \in \mathcal{B}$ a number $\epsilon > 0$, interval I and an increasing sequence of positive numbers (t_n) such that for all n

$$d_I(S_{t_n}(u), \mathcal{A}) \geq \epsilon .$$

Since $\mathcal{B}$ is a compact subset of $L^2(I, dx)$ (use Montel's theorem) we can find a subsequence (n_j) such that the sequence $S_{t_{n_j}}(u)$ converges in $L^2(I, dx)$ to a function v_*. Since the sequence of sets $S_{nT}(\mathcal{B})$ is decreasing it is easy to verify that $v_* \in \mathcal{A}$, a contradiction. This implies that $\mathcal{A}$ is a globally attracting set in all the d_I topologies.

With a little more work one can prove that $S_t(\mathcal{B})$ converges to $\mathcal{A}$ in the Hausdorff distance. Finally we mention two results about invariant sets.

Theorem III.3. *The set of almost periodic functions is invariant by the semiflow of CGL.*

We recall that a function u is almost periodic if it is uniformly continuous, and if for any $\epsilon > 0$ there is a number $\Lambda(\epsilon) > 0$ such that any interval of length $\Lambda(\epsilon) > 0$ contains a number λ (depending in general of ϵ) such that

$$\sup_x |u(x) - u(x + \lambda)| \leq \epsilon \, .$$

A more general space than the space of almost periodic functions was introduced by Wiener (Wiener 1967). A function u is said to belong to the space S' if for all real x the following limit exists

$$C_{u,u}(x) = \lim_{L \to \infty} \frac{1}{2L} \int_{-L}^{L} \overline{u}(x + y) u(y) \, dy \, ,$$

and is continuous in x.

Wiener defined and studied these functions and proved in particular that these functions have a well defined power spectrum (nowadays one would do it easily using Bochner's Theorem, see for example Chempeney, 1973). Note that these are also the functions for which the spatial correlations are well defined. In particular if we consider a measure μ on the function space which is translation invariant and ergodic for this group, then for μ almost every function the above limit exists (continuity of $C_{u,u}$ comes from the uniform continuity of u). In some sense this says that S' contains the measure theoretic support of μ. We therefore expect this space to be interesting if we want to study statistical properties of (CGL). The above argument using ergodicity can be generalized to any power of u, and this leads one to define the set S^∞ to be the set of uniformly continuous bounded functions such that for any integer $n \geq 1$ and any $x_1, \cdots, x_n$, the limits

$$\lim_{L \to \infty} \frac{1}{2L} \int_{-L}^{L} dy \prod_{j=1}^{n} u^{\#_j}(x_j + y)$$

exist, and are continuous functions of $x_1, \cdots, x_n$, where $u^{\#} = u$ or $\overline{u}$. From what we said above the following result can be expected.

Theorem III.3. *The set S^∞ is invariant by the semiflow of CGL.*

Theorems III.2 and III.3 follow from the local time existence Theorem. We leave the details to the reader.

4. Instability of the continuous spectrum

A very fruitful idea in the study of finite dimensional dynamical systems is the introduction of a parameter. By varying this parameter one can study the development of the complexity of the dynamical behaviour. Particularly important phenomena are bifurcations and transitions to chaos. In extended systems one can try to follow the same ideas, and we will explain them below on the particularly convenient example of the Swift Hohenberg (S.H.) equation (defined at the end of section II), referring to the literature for other examples.

One of the first steps when using a parameter in the study of dynamical systems is to locate first a parameter range where the system has a simple and well understood dynamical behaviour. For the (S.H.) equation, $u = 0$ is a stationary solution, and for $\eta < 0$, it is easy to verify that it is a linearly stable solution. Indeed, the linearized operator $\mathcal{L}$ is given by

$$\mathcal{L} = \eta - (1 + \partial_x^2)^2 ,$$

and using the Fourier transformation one checks that its spectrum is the set

$$\Sigma_\eta = \{\eta - (1 - k^2)^2 \mid k \in \mathbf{R}\}$$

which is contained (strictly) in the left half of the complex plane (in fact in the negative real line) for $\eta < 0$. Note that this is the spectrum in the space $L^2(\mathbf{R})$ which is not really what we are interested in. However it is left to the reader to prove that the spectrum is the same in $L^\infty(\mathbf{R})$. Using the local energy methods of the previous chapter, it is even possible to prove that $u = 0$ is a globally stable stationary solution, namely any bounded regular initial condition gives rise to a solution tending to zero in the supremum norm. For $\eta > 0$ however, the spectrum of the linearized operator has a part in the right half plane of the complex plane. Therefore we expect $\eta = 0$ to be somewhat analogous to a bifurcation point. A first difference however occurs immediately. In standard bifurcation theory of a stationary solution, it is a discrete spectrum which becomes unstable; only finitely many eigenvalues of the linearized evolution equation at the fixed point cross the imaginary axis at the bifurcation point. Here we have a continuous spectrum becoming unstable (Σ_η is a line in the complex plane). As we will see below this will create severe difficulties, but also we will see beyond the instability point a rather rich dynamical behaviour.

We mention that some particular questions can be studied using the standard techniques. A typical example is the occurrence of solutions periodic in space of fixed period. This would amount to fixing periodic boundary conditions on a finite interval. Then one can try to use the standard techniques to study bifurcations and other problems. However when it comes to stability questions they should be formulated in a global phase space, not in a space of periodic functions. We refer to Collet & Eckmann (1990a) for some examples.

A rather powerful tool in the study of bifurcations of a stationary solution is the method of normal forms, and its elementary version the amplitude equation. The idea is to try by rescaling and changes of variable to put the equation in a simplest form as possible, and then to study the dynamical behaviour of this simpler equation. An appreciable gain in this program is also that one often finds the same simple form, and this leads to a classification of the bifurcations. We are going to try to apply the same ideas for questions about extended domains. Of course the results are not as nice as for the standard theory of normal forms, and the technical difficulties are much worse. However as we will see below, this is a fruitful idea to pursue. Before we go on, a word of caution is needed. Even in finite dimensional dynamical systems, it is rarely true that a dynamical system is conjugate to a (truncated) normal form. One usually computes a normal form to a given order and discards the rest, assuming it to be small. This means that one cannot deduce rigorously the dynamical behaviour of a dynamical system from the dynamical behaviour of its truncated normal form. Such truncated normal forms give a good starting point to investigate the dynamics but one has to provide a separate proof for the control of any particular trajectory in the initial dynamical system. Another way to view this problem, often disregarded in the literature, is that the dynamics of the truncated normal form will describe the dynamics of the real system for a rather large time with a small error. However for very large time one usually loses control on the size of the error, and the normal form becomes useless. We refer to Il'yashenko & Yakovenko (1991) for more information. This problem will of course also be present in the extended system, and the method of amplitude equation will only claim to describe the dynamics accurately for a large but finite time.

There are two basic steps in applying the method of amplitude equations, both inspired by the finite dimensional experience. The first step comes from a general idea in bifurcation theory that the interesting quantity to look at is the amplitude of the solution in the direction becoming unstable. In finite dimensional systems this is easily determined since we have only finitely many eigenvalues becoming unstable, so we only have to use as coordinates the corresponding spectral subspace, and this leads to the theory of centre manifold. In our situation of the (S.H.) equation, we will consider what occurs exactly at the bifurcation point $\eta = 0$. We have then only the point zero of the spectrum on the imaginary axis, and to this point corresponds two (generalized) eigenvalues $e^{\pm ix}$. Therefore since we are only interested in real solutions, we will look for them in the form

$$u(x,t) = Be^{ix} + c.c.$$

In finite dimensional problems, the amplitude B is only a function of time, here of course we will require that B is a function of space and time. The second step of the amplitude analysis is now to fix adequately the scaling for the space and time dependence of B. It is only for some adequate scaling that we will obtain an interesting non trivial equation for B. The basic idea is roughly to try to find a scaling such that in the equation all terms will be of the same order. Guessing the correct scaling may be sometimes difficult, we refer to Maslov (1976) for more information on this problem. Here however the situation is quite simple looking first at the linearized operator for $\eta > 0$ small. It is then a good idea to introduce the square root of η as a parameter, namely to set $\eta = \epsilon^2$ (we will take $\epsilon > 0$ for definiteness). In Fourier space, the evolution linearized around $u = 0$ reads

$$\partial_t v(k,t) = \omega(\epsilon,k) v(k,t)$$

with

$$\omega(\epsilon,k) = \epsilon^2 - (1-k^2)^2 \, .$$

We therefore obtain two scales for k. The first one $|k| \approx 1$ corresponds to the two generalized eigenvalues $e^{\pm ix}$ that we have already introduced. There is however another scale for k which is obtained by looking at the behaviour of $\omega(\epsilon,k)$ near $k = \pm 1$. Let us set for example $k = 1 + q$ with q small, one gets

$$\omega(\epsilon, 1+q) = \epsilon^2 - 4q^2 + \mathcal{O}(q^3) \, ,$$

and we derive a natural scale for q which is $|q| \approx \epsilon$. Having two scales for k we derive two scales for the dual variable x, namely the scale 1 and a large scale $1/\epsilon$. Since we have several scales the method is often called the multi-scale method. At the scale $|q| \approx \epsilon$ we have $\omega \approx \epsilon^2$ and we get a natural time scale ϵ^{-2}. Finally in order to equilibrate also the non linear term (otherwise we get a true but uninteresting trivial result) it is natural to use a scale ϵ for the amplitude of B. In other words we are lead to the ansatz

$$u(x,t) = \epsilon e^{ix} \mathcal{A}(\epsilon x, \epsilon^2 t) + c.c.$$

Note that such functions can be interpreted as large scale (in space and time) modulations of the basic function e^{ix}. It is in this sense that the above ansatz will be interesting and non trivial, after all any real function of x and t can be written in this form. In order to see that we have indeed made the correct guess for the functional form of the ansatz, we have to put this ansatz into the (S.H.) equation. After some algebra one gets the following equation for the function $\mathcal{A}(y,\tau)$

$$0 = -\partial_\tau \mathcal{A} + 4\partial_y^2 \mathcal{A} + \mathcal{A} - 3\mathcal{A}|\mathcal{A}|^2 - e^{2iy/\epsilon}\mathcal{A}^3 + 4i\epsilon\partial_y^3 \mathcal{A} - \epsilon^2 \partial_y^4 \mathcal{A} + c.c.$$

It would then be nice to be able to drop the terms $4i\epsilon\partial_y^3 \mathcal{A} - \epsilon^2\partial_y^4 \mathcal{A}$ because they are small and the term $e^{2iy/\epsilon}\mathcal{A}^3$ because it is rapidly oscillating and should in some sense average to a small quantity. If we do this and introduce for convenience the function $A(x,t) = \sqrt{3}\mathcal{A}(2x,t)$ we get the Ginzburg-Landau equation

$$\partial_t A = \partial_x^2 A + A - A|A|^2 \,. \tag{GL}$$

In other words, the Ginzburg-Landau equation appears as a truncated amplitude equation for the instability of the zero solution of the Swift Hohenberg equation.

As explained above this is not at all the end of the story, in particular we have to prove that the three terms we have neglected to obtain the (G.L.) equation are indeed small and that they do not overly upset the dynamical behaviour for some large time.

In order to formulate the result we will denote by GL_t the (non-linear) semi-group of evolution for the (GL) equation and by SH_t the corresponding semi group for the Swift Hohenberg equation.

Theorem IV.1. *For any $T > 0$ there is an $\epsilon_0 > 0$ and a constant $C > 0$ such that for any $\epsilon \in]0,\epsilon_0]$, for any $t \in [0, T/\epsilon^2]$ and for any $A_0 \in C_b^{4,u}$, we have with $u_0(x) = 3^{-1/2}\epsilon A_0(\epsilon x/2)e^{ix} + c.c.$*

$$\sup_x \left| SH_t(u_0)(x) - \epsilon 3^{-1/2} e^{ix} GL_{t\epsilon^2}(A_0)(\epsilon x/2) - c.c. \right| \leq C\epsilon^3 \min(t^{1/2}, t^{3/4}) \,.$$

We now make some remarks about this result. As explained before this result only says that we have some control of the solution of the (SH) equation by the (GL) equation for some large but finite time, and one cannot expect more.

Several versions of this Theorem have been published using different spaces. As explained before the most interesting spaces are those with no conditions at infinity. The above formulation first appeared in Collet & Eckmann (1990b). We refer to Kirrmann *et al.* (1992) for some extensions (in particular of the time interval), and to van Harten (1991) for another proof.

There is another rather unsatisfactory point about this result which is that it assumes a special form for the initial data u_0. This problem was discussed by Eckhaus (1993) who proved that starting from any regular bounded initial data u_0 for (SH), after a time of order $\log \epsilon^{-1}$, the solution has the required functional form and one can apply the above Theorem. Note that it takes a relatively small amount of time for the solution to develop the correct form. As a consequence, after a time interval T/ϵ^2 where the Theorem does not guarantee anymore a good approximation of the (SH) evolution, one can simply apply the Theorem again with another A_0, obtaining this way a chain of good approximations (see Kirrmann *et al.* (1992) for more details).

We also observe that in the bound $\min(t^{1/2}, t^{3/4})$ the first term $t^{1/2}$ dominates at short times. This is coming from the fact that in the initial regime, the functions cannot be assumed very regular, and estimates on the regularity only develop and stabilize as time goes on (see the previous section for a similar result).

We now give some ideas about the proof of Theorem IV.1. Let $C(x,t)$ be a solution of

$$\partial_t C = \partial_x^2 C + C - C|C|^2 - \frac{1}{3} e^{2ix/\epsilon} C^3 + \frac{i\epsilon}{2} \partial_x^3 C - \frac{\epsilon^2}{16} \partial_x^4 C ,$$

and let A be a solution of (GL)

$$\partial_t A = \partial_x^2 A + A - A|A|^2$$

both with initial data $A_0(x) = A(x,0) = C(x,0)$. We want to estimate $w = C - A$. Writing $C = A + w$, we obtain an equation for w which has the following form

$$\partial_t w = Lw + F + NL(w)$$

where F is a term independent of w given by

$$F = -\frac{1}{3} e^{2ix/\epsilon} A^3 + \frac{i\epsilon}{2} \partial_x^3 A - \frac{\epsilon^2}{16} \partial_x^4 A ,$$

L is a linear operator given by

$$Lw = \partial_x^2 w + w + \frac{i\epsilon}{2} \partial_x^3 w - \frac{\epsilon^2}{16} \partial_x^4 w - 2|A|^2 w - A^2 \overline{w} - e^{2ix/\epsilon} A^2 w ,$$

and $NL(w)$ is a polynomial in w and $\overline{w}$ of degree 3 with no constant or linear terms. Note that L is not really linear because of the presence of $\overline{w}$, and one should in principle introduce the evolution equation for the (real) vector of real and imaginary parts of w. To avoid some heavy notations we will continue as if L was linear.

We can immediately make two simple remarks. First, since A is regular (see Section II), the last two terms in F are small. Second, as long as w remains small, we can in a first approach neglect $NL(w)$. Note that by construction, $w(x,0) = 0$. The whole proof is in fact based on a contraction mapping principle, and we will below consider only the affine part of the evolution equation for w, referring to the original papers for the complete proof. There are basically two difficulties, namely L is an operator with variable coefficients (some of them varying rather wildly like $e^{2ix/\epsilon}A^2$); the second difficulty is that the first term in F is not obviously small, and some sort of averaging will have to be performed. The first basic step is to try to control the kernel of the semi-group generated by L. For $t \geq \tau$, let $G(x,t;y,\tau)$ be the function such that $G(x,t;y,t) = \delta(x-y)$, and for $t > \tau$

$$\partial_t G(x,t;y,\tau) = L_{x,t} G(x,t;y,\tau) \ .$$

With this function, we can solve the equation

$$\partial_t w = Lw + F$$

with initial data $w(x,0) = w_0(x)$, namely

$$w(x,t) = \int G(x,t;y,0) w_0(y) dy + \int_0^t ds \int G(x,t;y,s) F(y,s) dy \ .$$

If we assume $w_0 = 0$, we only have to control the second term. Assume that G is known to have an L^1 norm in y bounded in x, then this is enough to control (the L^∞ norm of) the contribution to w of the last two terms of F which are of order ϵ. Assume also that $|\partial_y G|$ is also integrable in y and that this integral is integrable in τ with a uniform bound in x. We cannot expect much more to be true since for the heat kernel $G^H(x,t;y,\tau)$ we have

$$\partial_y G^H(x,t;y,\tau) = \frac{2(x-y)}{\sqrt{4\pi(t-\tau)}(t-\tau)} e^{-(x-y)^2/4(t-\tau)} \ .$$

The L^∞ norm in y of $\partial_y G^H(x,t;y,\tau)$ is of order $1/(t-\tau)$ which diverges when τ increases to t and is not integrable in τ. However the L^1 norm with respect to y of this singularity is integrable in τ. We can now use integration by parts (noticing that $G(x,t;\pm\infty,\tau) = 0$) to estimate the contribution of the first term of F, namely

$$\int_0^t ds \int G(x,t;y,s) e^{2iy/\epsilon} A^3(y,s) dy$$

$$= \int_0^t ds \int G(x,t;y,s) A^3(y,s) d\left(\frac{\epsilon e^{2iy/\epsilon}}{2i}\right)$$

$$= \frac{i\epsilon}{2} \int_0^t ds \int \partial_y \left[G(x,t;y,s) A^3(y,s) \right] e^{2iy/\epsilon} dy \ ,$$

and the bound follows easily.

In order to prove the above assumptions on $G(x,t;y,s)$, we write

$$L = L_1 + L_2$$

with

$$L_1 = \partial_x^2 + 1 + \frac{i\epsilon}{2}\partial_x^3 - \frac{\epsilon^2}{16}\partial_x^4 \ .$$

Since this is an operator with constant coefficients we can use Fourier analysis to control the kernel of the associated semigroup $G^1(x,t;y,s)$. One obtains easily by direct estimates the existence of a constant $D > 0$ such that for any $t > 0$ and for $r = 0,\ 1$

$$\int |\partial y^r G^1(x,t;y,\tau)| dy \le D \min((t-\tau)^{-r/2}, (\epsilon^2(t-\tau))^{-r/4}) \ .$$

The operator L_2 is given by

$$L_2 w = -2|A|^2 w - A^2 \overline{w} - e^{2ix/\epsilon} A^2 w \ ,$$

and as explained above, we will proceed as if this was a linear operator. Using Dirac's formula we have

$$G(x,t;y,\tau) = G^1(x,t;y,\tau) - \int_\tau^t \int G(x,t;z,s) L_2(z,s) G_1(z,s;y,\tau) ds\, dz$$

$$= G^1(x,t;y,\tau) + \int_\tau^t \int G^1(x,t;z,s) L_2(z,s) G(z,s;y,\tau) ds\, dz \ .$$

These are inhomogeneous equations for G which can be solved iteratively for $t-\tau$ small. The first equation is used to control the operator G acting in L^∞. The second one is used to control the operator $\partial_y G$. We refer to Collet & Eckmann (1990b) for the complete details.

5. Large time behaviour

In this section we are going to study the large time behaviour of some interesting solutions of the Ginzburg Landau equation in dimension one. We first observe that this solution possesses a one parameter family of stationary solutions given by

$$A_q(x) = \sqrt{1-q^2} e^{iqx} \qquad \text{for} \quad q \in [-1,1] \ .$$

It is well known (Collet & Eckmann, 1990a) that these solutions are unstable for $|q| > 1/\sqrt{3}$ and are marginally stable for $|q| \leq 1/\sqrt{3}$ (the Eckhaus instability). In other words, for $|q| \leq 1/\sqrt{3}$ the spectrum of the linearized operator is contained in the closed left half plane and tangent to the imaginary axis. For simplicity we will refer to these solutions as stable in what follows.

Some interesting initial conditions are obtained by concatenating two such stationary solutions, one near $+\infty$, and another one near $-\infty$. Namely for a number $L > 0$ and $q_\pm \in [-1, 1]$ we define an initial condition by

$$A_i(x) = \begin{cases} A_{q_-}(x) & \text{for } x < -L, \\ A_{q_+}(x) & \text{for } x > L, \end{cases}$$

and some regular interpolation between the two stationary solutions for $-L \leq x \leq L$. If one of the asymptotic stationary solution is stable and the other one unstable, the intuitive picture is that in the evolution the stable solution should invade the unstable one. This is indeed what happens, and this problem leads to the existence of front solutions. These are solutions which interpolate between the two asymptotic stationary states and move with a definite velocity without deformation (or at least with small deformation). We refer to Collet & Eckmann (1990a) and Gallay (1994) for more information and references to the literature.

We will consider below the case of two stable solutions, namely $|q_\pm| \leq 1/\sqrt{3}$. In this case the intuition about the large time behaviour is not so clear. Note that in the space $\mathbf{R} \times \mathbf{C}$ the graph of the stationary solution $\{(x, A_q(x))\}$ looks like a spiral. One possible scenario of the evolution is that one solution invades the other by "expansion" or "compression" of the spiral at the interface. Another possible scenario is that one spiral would loose (or gain) some turn by passing through zero (in the complex plane). This is called a phase slip. We will see below that none of these simple mechanisms is the right one.

There is a useful remark to be made about solutions which have no zero, namely that they can be represented by amplitude and phase functions, $R > 0$ and φ, as

$$A(x,t) = R(x,t)e^{i\varphi(x,t)} ,$$

and as long as $A \neq 0$ we have the following system of equations for R and φ which is equivalent to (GL)

$$\partial_t R = R'' + R - R\varphi'^2 - R^3 ,$$

$$\partial_t \varphi = \varphi'' + 2\frac{R'}{R}\varphi' . \tag{V.1}$$

Note that although this system is singular when $R = 0$, this singularity is only a singularity of the representation of the solution in amplitude and phase, not of the solution of the Ginzburg Landau equation.

We can now formulate the first result which says that for suitable initial conditions phase slips do not occur.

Theorem V.1. *There is a number* $\epsilon_0 > 0$ *such that if the function*

$$A_0(x) = R_0(x)e^{i\varphi_0(x)}$$

satisfies

$$\|R_0'\|_{H^1} + \|\varphi_0'\|_{H^1} + \|R_0^3 - R_0 + R_0{\varphi_0'}^2\|_{H^2} \leq \epsilon_0$$

then the solution $A(x,t)$ *of the Ginzburg Landau equation with initial data* A_0 *satisfies for any* $t > 0$

$$\inf_x |A(x,t)| > 0 \ .$$

In fact one can even derive a uniform positive lower estimate for the modulus of A. Note that the hypothesis says that the solution looks everywhere locally like a stationary solution. Note also that we have not assumed convergence at $\pm\infty$ to a stationary solution, therefore this result is more general than what is required to study our special initial condition. We refer to Collet & Eckmann (1992) for more details.

The system of equations (V.1) is not well adapted to making estimates, and a more useful system is obtained by introducing the local wave number $\eta = \varphi'$, for which we obtain the new system

$$\partial_t R = R'' + R - R\eta^2 - R^3 \ ,$$

$$\partial_t \eta = \eta'' + 2(R'\eta/R)' \ .$$

We will only consider small wave numbers below; for stationary solutions this gives amplitude near 1 and it is therefore interesting to introduce the (small) quantity $s = R - 1$. We then obtain the system of equations

$$\partial_t s = s'' - 2s - \eta^2 - s\eta^2 - 3s^2 - s^3 \ ,$$

$$\partial_t \eta = \eta'' + 2(s'\eta/(1+s))' \ .$$

There are several interesting approximations to this system. First, in the second equation, if we neglect the second term (which is quadratic), we obtain the diffusion equation for the local wave number η. If we consider the first equation, for small s and η the function s should be in some sense slaved to the forcing term $-\eta^2$, namely we should have to a good approximation $s = -\eta^2/2$. Note that this corresponds to the first term in the development of the amplitude of the stationary solution near zero wave number. Introducing this expression in the evolution equation for η, we get the equation

$$\partial_t \eta = \partial_x \left(\frac{2 - 3\eta^2}{2 - \eta^2} \partial_x \eta \right) ,$$

which is similar to some porous medium equations.

In order to understand the large time behaviour of our problem we are going to consider first the simplest approximation, namely the heat equation for the local wave number. Note that if we write the initial condition in terms of amplitude and phase, we get that the amplitude and local wave number have limits at $\pm\infty$.

Theorem V.2. *Let $\eta_0(x)$ be a uniformly continuous function on* $\mathbf{R}$ *with limits at $x = \pm\infty$. Then the solution of the heat equation with initial condition η_0 converges to the number $(\eta_0(-\infty) + \eta_0(+\infty))/2$ when $t \to \infty$ uniformly on compact sets.*

The proof is an easy consequence of the explicit formula giving $\eta(x, t)$ in terms of the heat kernel and η_0. The details are left to the reader.

Note that it is important to assume that the initial condition has limits at $\pm\infty$, we refer to Collet & Eckmann (1992) for some counter examples if this assumption is not satisfied and to Denisov & Repnikov (1984) for a general condition of convergence. Also the convergence is certainly not uniform over the whole line. At time $t > 0$ and at a distance much larger than $\sqrt{t}$ from the origin, the solution has essentially retained its initial value.

The next step of difficulty would be to investigate the large time evolution of the porous medium equation derived above. We shall first investigate a simpler problem. The porous medium equation corresponds to a heat diffusion equation with a variable diffusion coefficient. We will assume for a moment that this coefficient $D(x)$ is time independent (and independent of η) and study the large time behaviour of the parabolic linear equation

$$\partial_t \eta = \partial_x \left(D(x) \partial_x \eta \right) . \tag{V.2}$$

Note that on the right hand side appears the gradient of the heat flux and this flux is proportional to the gradient of temperature. We will assume that the function D is uniformly regular, strictly positive and has limits $D_\pm > 0$ at $x = \pm\infty$.

Since this is a problem with non constant coefficients, the standard Fourier techniques cannot be used. We will first start with the simpler case where D is a step function, namely

$$D^s(x) = \begin{cases} D_- & \text{if } x < 0, \\ D_+ & \text{if } x > 0, \end{cases}$$

where D_+ and D_- are two different positive numbers. The initial condition η_0 is chosen as in Theorem V.2.

Proposition V.3. *The solution of the equation*

$$\partial_t \eta = \partial_x \left(D^s(x) \partial_x \eta \right) , \tag{V.3}$$

with initial condition η_0 as in Theorem V.2 converges to the number

$$\frac{\sqrt{D_-}\, \eta_0(-\infty) + \sqrt{D_+}\, \eta_0(+\infty))}{\sqrt{D_-} + \sqrt{D_+}}$$

when $t \to \infty$ uniformly on compact sets.

Note that contrary to what was said before, the function $D^s(x)$ is not regular. In the above equation, it is assumed that the solution is such that the flux $D^s(x)\partial_x \eta(x,t)$ is continuous in x.

Proof. The first step is to introduce a special solution which deals with the behaviour at infinity and which has some special scaling properties. Such solutions are often called scaling solutions. They can often be guessed with the use of dimensional analysis. Here we will look for a solution of the form $\eta(x,t) = S(x/\sqrt{t})$ having the same limits as η_0 at infinity. In other words, introducing the variable $y = x/\sqrt{t}$ we have to solve the equation

$$\frac{-yS'(y)}{2} = \left(D^s(y) S'(y) \right)' ,$$

with the conditions at infinity $y(\pm\infty) = \eta_0(\pm\infty)$. There is of course a problem with this equation for $y = 0$. We first consider the half line $y > 0$, where it is easy to integrate the equation, and the general solution is given by

$$S(y) = B_+ \int_0^y e^{-z^2/4D_+} dz + A_+ ,$$

where B_+ and A_+ are two constants. Similarly, for $y < 0$ one gets

$$S(y) = B_- \int_0^y e^{-z^2/4D_+} dz + A_- .$$

We now have to determine the four constants $A_\pm$ and $B_\pm$. We have two conditions at infinity, namely $S(\pm\infty) = \eta_0(\pm\infty)$. We will of course impose S to be continuous at the origin $S(0_-) = S(0_+)$, and as explained above we will impose that the heat flux is continuous at the origin, namely $D_- S'(0_-) = D_+ S'(0_+)$. We now have four linear relations and four unknown, and we leave to the reader the exercise of verifying that the linear system can indeed be solved. The most interesting information is the value of $S(0)$ which turns out to be

$$S(0) = \frac{\sqrt{D_-}\,\eta_0(-\infty) + \sqrt{D_+}\,\eta_0(+\infty))}{\sqrt{D_-} + \sqrt{D_+}} \,.$$

So far we have achieved the following. Let $a > 0$, and assume that we have an initial condition $\eta(x,0) = S(x/\sqrt{a})$. Then the solution of our equation is exactly $S(x/\sqrt{a+t})$. It is obvious form the expression of S that when $t \to +\infty$ this function (of x) converges to the the number $S(0)$ uniformly on compact sets.

We now consider a general initial condition η_0 (satisfying the assumptions of Theorem V.2). Let w_0 be the function defined by

$$w_0(x) = \eta_0(x) - S(x) \,.$$

Similarly, let $w(x,t)$ be defined by

$$w(x,t) = \eta(x,t) - S(x/\sqrt{t+1}) \,.$$

Since the heat equation (V.3) is linear, w satisfies also this equation, but the initial condition w_0 is uniformly continuous and tends to zero at $\pm\infty$. We would like of course to prove that w tends to zero when $t \to \infty$. This follows immediately from a deep theorem of Nash & Aronson, namely there is a constant $c > 1$ (depending only on $D_\pm$) and a function $K(x,t;y,\tau)$ defined for $t > \tau$ such that

$$w(x,t) = \int K(x,t;y,0) w_0(y) dy \,,$$

and

$$\frac{c^{-1}}{\sqrt{t-\tau}} e^{-c(x-y)^2/(t-\tau)} \le K(x,t;y,\tau) \le \frac{c}{\sqrt{t-\tau}} e^{-c^{-1}(x-y)^2/(t-\tau)} \,.$$

We refer to (Stroock 1988) for a proof and some more results. Using only the upper bound in the above formula, we leave to the reader to verify that $w(x,t)$ converges to zero uniformly on compact sets. This finishes the proof of the Proposition.

We are now going to discuss the same asymptotic problem with an x dependent regular diffusion coefficient $D(x)$ as introduced above in (V.2). This can be done by purely functional analytic methods, but we will instead introduce the very powerful technique of renormalization. This technique was developed recently by Bricmont & Kupiainen (1992) in the context of extended systems and has led to the solution of several important open problems. It uses scaling arguments developed by Barenblatt (1979) and Goldenfeld *et al.* (1990). In order to determine the large time behaviour of the solution η of (V.2) with initial condition η_0, we are going to proceed in time by exponential steps. To do so, we first fix a number $L > 1$ which is going to be used as a scaling factor. We now assume that the equation has been integrated on the time interval $[0, 1]$, so that we know the solution $\eta(x, t)$ for $0 \leq t \leq 1$. Knowing in particular $\eta(x, 1)$ we can imagine integrating the equation (by whatever method) on the time interval $[1, L^2]$. Again knowing $\eta(x, L^2)$ we can imagine integrating on the time interval $[L^2, L^4]$ and so on for every time interval $[L^{2n}, L^{2n+2}]$. Up to now this does not seem very interesting. In particular the process of integrating on time intervals amounts essentially to solving the initial problem. We will see however that later on a simpler and more tractable question will emerge from these considerations. We can consider that the integration on the time interval $[1, L^2]$ is an operator $\mathcal{T}_1$ which maps the data $\eta(x, 1)$ to the function of two variables $\eta(x, t)$ for $1 < t \leq L^2$. Similarly, we can introduce an operator $\mathcal{T}_2$ which maps the data $\eta(x, L^2)$ to the function of two variables $\eta(x, t)$ for $L^2 < t \leq L^4$. The main fact is that there is some relation between $\mathcal{T}_1$ and $\mathcal{T}_2$. To see this we are going to scale down the solution on the time interval $[L^2, L^4]$. Define for $1 \leq t \leq L^2$ the function v by

$$v(y, \tau) = \eta(Ly, L^2\tau) \; ,$$

and let us determine the equation for $v(y, \tau)$ on the time interval $1 \leq \tau \leq L^2$. One easily gets

$$\partial_\tau v = \partial_y \left(D(L \,\cdot) \partial_y v \right) \; .$$

Except for the change of scale in the diffusion coefficient, this is the same as equation (V.2), in other words we have a relation between the two operators $\mathcal{T}_1$ and $\mathcal{T}_2$. We can now build a more systematic approach by considering the operator $\mathcal{T}_1$ as a function of the diffusion coefficient, we will denote it by $\mathcal{T}(D)$. We have just seen that

$$\mathcal{T}_2 = \mathcal{S}_L^{-1} \circ \mathcal{T}(D(L \,\cdot)) \circ \mathcal{S}_L \; ,$$

where $\mathcal{S}_L$ denotes the scaling operator

$$\mathcal{S}_L(u)(x, t) = u(Lx, L^2 t) \; .$$

Since we are also scaling the solution, it is more adequate to consider simultaneously the pair (η, D), and more generally the couple made of the solution (on the time interval $[1, L^2]$), and the equation. The renormalization operator $\mathcal{R}_L$ is now defined as acting on the pairs (η, D) by

$$\mathcal{R}_L(\eta, D)(x,t) = (\eta(Lx, L^2t), D(Lx)) = (\mathcal{T}(D(L\,\cdot))(\mathcal{S}_L\eta)(x,t)\,,\, D(Lx))\,.$$

We can of course now iterate the the renormalization operator. In particular

$$\mathcal{R}_L^n(\eta, D)(x,t) = (\eta(L^n x, L^{2n}t), D(L^n x))\,.$$

The main outcome of this formula (which at first seems rather trivial) is that if we somehow know the left hand side, we know the solution η at large time (and at large scale). Since the left hand side corresponds to the iteration of the operator $\mathcal{R}_L$, it is natural to look for a fixed point and determine its stability. A fixed point (η^*, D^*) of $\mathcal{R}_L$ satisfies $\mathcal{R}_L(\eta^*, D^*) = (\eta^*, D^*)$. In particular $D^*(Lx) = D^*(x)$, and we immediately recognize that our function $D^s(x)$ is such a fixed point. Similarly, the function η^* should satisfy (V.3) and $\eta^*(Lx, L^2t) = \eta^*(x,t)$, and we verify easily that our function $S(x/\sqrt{t})$ satisfies these two requirements. In other words, we start with the function $\eta^*(x,t) = S(x/\sqrt{t})$ on the interval $1 \le t \le L^2$, construct the function $v(y,1)$ by

$$v(y,1) = \eta^*(Ly, L^2)\,,$$

which happens to be equal to the function $S(y)$. We then solve (V.3) with this initial data up to time L^2 obtaining the function $S(y/\sqrt{t})$ which is indeed our starting point. The stability of this fixed point can be checked in suitable function spaces (see Bricmont-Kupiainen 1992) and this leads to the following result.

Theorem V.4. *Let $D(x)$ be a regular uniformly continuous function, strictly positive with non zero limits $D_\pm$ at $\pm\infty$. Assume also that $(D - D_+)^2$ is integrable near $+\infty$, and $(D - D_-)^2$ is integrable near $-\infty$. The solution of the equation*

$$\partial_t \eta = \partial_x \left(D(x)\partial_x \eta\right)\,,$$

with initial condition η_0 as in Theorem V.2 and such that $\eta_0 - \eta^*$ belongs to L^2 converges to the number

$$\frac{\sqrt{D_-}\,\eta_0(-\infty) + \sqrt{D_+}\,\eta_0(+\infty))}{\sqrt{D_-} + \sqrt{D_+}}$$

when $t \to +\infty$ uniformly on compact sets.

Proof. We are first going to prove convergence in L^2 norm. As advocated above, we define the function $\eta_n(x,t)$ for $1 \le t \le L^2$ by

$$\eta_n(x,t) = \eta(L^n x, L^{2n}t)\,.$$

Note that this function satisfies the equation

$$\partial_t \eta_n = \partial_x D_n \partial_x \eta_n \tag{V.4}$$

where $D_n(x)$ denotes the function $D(L^n x)$. We are then going to consider on the time interval $[1, L^2]$ the function

$$Z_n(t) = \int (\eta_n(x,t) - \eta^*(x,t))^2 \, dx \ ,$$

which is a measure of the distance between η_n and η^* We have easily using integration by parts, (V.3) and (V.4)

$$\frac{d}{dt} Z_n(t) = -2 \int D_n(x) \left(\partial_x \eta_n(x,t) - \partial_x \eta^*(x,t)\right)^2 dx$$

$$-2 \int (D_n(x) - D^s(x))(\partial_x \eta_n(x,t) - \partial_x \eta^*(x,t)) \partial_x \eta^*(x,t) dx \ ,$$

and by completing the square we finally get

$$\frac{d}{dt} Z_n(t) \le - \int D_n(x) \left(\partial_x \eta_n(x,t) - \partial_x \eta^*(x,t)\right)^2 dx$$

$$+ \int (D_n(x) - D^s(x))^2 \frac{(\partial_x \eta^*(x,t))^2}{D_n(x)} dx \ .$$

Integrating from $t = 1$ to $t = L^2$, and denoting by $\delta > 0$ a lower bound on the function D, we get

$$Z_n(L^2) - Z_n(1) \le$$

$$\int (D_n(x) - D^s(x))^2 g(x) dx - \int_1^{L^2} ds \int D_n(x) \left(\partial_x \eta_n(x,s) - \partial_x \eta^*(x,s)\right)^2 dx \ , \tag{V.5}$$

where

$$g(x) = \delta^{-1} \int_1^{L^2} (\partial_x \eta^*(x,s))^2 ds.$$

We leave to the reader to verify that the positive function g is integrable in x. The key observation is now that

$$Z_n(L^2) = L Z_{n+1}(1) \ .$$

We therefore get the following recursive estimate

$$Z_{n+1}(1) \le L^{-1} Z_n(1) + L^{-1} \int (D_n(x) - D^s(x))^2 g(x) dx \ .$$

Notice that we have obtained this inequality (which as we will see below tells us something useful about the time evolution) without a complete integration of the heat equation. It is important to realize at this point that the so-called renormalization group approach has lead us to prove the above inequality which is a simpler task than computing the kernel of the semi-group.

It follows from Lebesgue's dominated convergence theorem that the last term in the above expression goes to zero when n tends to infinity (since D_n converges to $D^* = D^s$ except at the origin). This immediately implies (since $Z_n(1) \geq 0$) that $Z_n(1)$ converges to zero when n tends to infinity. We again leave to the reader to prove the more precise estimate

$$Z_n(t) \leq CnL^{-n} \quad \text{for} \quad 1 \leq t \leq L^2$$

where C is a positive constant (this follows from the fact that the first term on the right hand side of (V.5) is bounded by $\mathcal{O}(1)L^{-n}$). In other words, we have proven that η_n converges to η^* in $L^2(\mathbf{R}, dx)$.

We now work to obtain convergence in a finer topology. First of all it follows at once from (V.5) and the previous estimate on $Z_n(1)$ that

$$\int_1^{L^2} ds \int D_n(x) \left(\partial_x \eta_n(x,s) - \partial_x \eta^*(x,s)\right)^2 dx \leq C' n L^{-n} ,$$

where C' is some positive constant. This implies that on the time interval $[1, C'nL^{-n}]$ (assuming n large enough so that $C'nL^{-n} < L^2$), there is a time $t_n \in [1, L^2[$ such that

$$\int D_n(x) \left(\partial_x \eta_n(x,t_n) - \partial_x \eta^*(x,t_n)\right)^2 dx \leq 1 .$$

Since

$$\int D_n(x) \left(\partial_x \eta^*(x,t)\right)^2 dx \leq \|D\|_\infty \int \left(\partial_x \eta^*(x,t)\right)^2 dx$$

does not depend on n, we see that there is a constant $c_1 > 0$ such that for any n large enough we have

$$\int D_n(x) (\partial_x \eta_n(x,t_n))^2 dx \leq c_1 .$$

We now observe that it follows easily from (V.4) that

$$\frac{d}{dt} \int D_n (\partial_x \eta_n)^2 dx \leq 0 .$$

Therefore on the whole interval $[t_n, L^2]$ we have

$$\int D_n(x) \left(\partial_x \eta_n(x,t)\right)^2 dx \leq c_1 .$$

This implies that there is a constant $c_2 > 1$ such that for any n large enough we have for t in $[t_n, L^2]$

$$\int D_n(x)\left(\partial_x\eta_n(x,t) - \partial_x\eta^*(x,t)\right)^2 dx \le c_2 .$$

Notice that we also get recursively an estimate for the missing time interval $[1, t_n]$ (if of course $t_n > 1$ otherwise we are done). Indeed, we observe that for $n > 1$

$$\int D_{n-1}(x)\left(\partial_x\eta_{n-1}(x,L^2) - \partial_x\eta^*(x,L^2)\right)^2 dx$$

$$= L^{-1}\int D_n(x)\left(\partial_x\eta_n(x,1) - \partial_x\eta^*(x,1)\right)^2 dx .$$

Therefore for n large enough we have

$$\int D_n(x)\left(\partial_x\eta_n(x,1) - \partial_x\eta^*(x,1)\right)^2 dx \;\le c_2 L ,$$

which implies by the same chain of arguments as above that there is a constant $c_3 > 1$ such that for any n large enough we have on $[1, t_n]$

$$\int D_n(x)\left(\partial_x\eta_n(x,t) - \partial_x\eta^*(x,t)\right)^2 dx \le c_3 .$$

Using that $D_n > \delta > 0$ for all n, and applying a standard Sobolev inequality we get finally

$$\sup_{t\in[1,L^2]}\ \sup_x |\eta_n(x,t) - \eta^*(x,t)| \le \mathcal{O}(1)n^{1/2}L^{-n/2} .$$

In other words, (η_n) converges to η^* in the sup norm and this immediately implies the claim since as we saw already, $\eta^*(x,t)$ converges uniformly on compact sets to the number $S(0)$.

This approach has been generalized in Bricmont & Kupiainen (1992) to the full system (V.1), and the following result was proved.

Theorem V.5. *For every initial condition A_0 sufficiently close to a "scaling solution", there is a stationary solution*

$$A^*(x) = \sqrt{(1-q^{*2})}e^{iq^*x+i\phi^*}$$

and a number θ^* such that the solution $A(x,t)$ of (GL) with initial data A_0 satisfies for any interval I

$$\lim_{t\to\infty}\sup_{x\in I}\left|A(x,t) - e^{i\theta^*\sqrt{t}}A^*(x)\right| = 0 .$$

In other words, a third stationary phase is nucleated and invades the whole space while slowly turning. Note that in general one cannot give an explicit formula for the new wave number q^*. Note also that there is no real scaling solution in this problem but only an approximate one. The proof is more intricate than the one given above for the heat equation but follows essentially the same steps. We refer to Bricmont & Kupiainen (1992) for a description of the topology and a complete proof.

References

Barenblatt, G. (1979) *Similarity, Self-similarity and Intermediate Asymptotics*. Consultants Bureau, New York.

Bricmont, J. & Kupiainen, A. (1992) Renormalization group and the Ginzburg-Landau equation. *Commun. Math. Phys.* **150**, pp. 193–208. See also Renormalizing Partial Differential Equations, to appear in the proceedings of the Mathematical Physics conference, Paris 1994.

Chempeney, D.C. (1973) *Fourier Transforms and their Physical Applications*. Acad. Press, London.

Collet, P. (1994a) Thermodynamic limit of the Ginzburg-Landau equation. *Nonlinearity* **7**, 1175-1190.

Collet, P. (1994b) Non linear parabolic evolutions in unbounded domains. In *Dynamics, Bifurcations and Symmetries*, pp. 97–104 in Chossat, P. (ed.), NATO ASI 437, Plenum, New York, London.

Collet, P. & Eckmann, J.-P. (1990a) *Instabilities and Fronts in Extended Systems*. Princeton University press, Princeton.

Collet, P. & Eckmann, J.-P. (1990b) The time-dependent amplitude equation for the Swift-Hohenberg problem, *Commun. Math. Phys.* **132**, pp. 139–153.

Collet P. & Eckmann, J.-P. (1992) Space-Time Behaviour in Problems of Hydrodynamic Type: A Case Study, *Nonlinearity* **5**, pp. 1265–1302.

Denisov, V. & Repnikov, V. (1984) The stabilization of a solution of a Cauchy problem for parabolic equations, *Differentsial'nye Uravneniya* **20**, pp. 20–41 (1984). *Differential equations* **20**, pp. 16–33.

Eckhaus, W. (1993) The Ginzburg-Landau equation is an attractor, *J. Nonlinear Science* **3**, pp. 329–348.

Friedman, A. (1964) *Partial Differential Equations of Parabolic Type*. Prentice Hall, Englewood Cliffs N.J.

Gallay, T. (1994) Local stability of critical fronts in nonlinear parabolic partial differential equations, *Nonlinearity* **7**, pp. 741–764.

Ginibre, J. & Velo, G. (1995) The Cauchy problem in local spaces for the complex Ginzburg Landau equation. I. Compactness methods. Preprint 95-76 LPTHE Orsay.

Goldenfeld, N., Martin, O., Oono, Y. & Liu, F. (1990) Anomalous dimensions and the renormalization group in a nonlinear diffusion process, *Phys. Rev. Lett.* **64**, pp. 1361–1364.

Hille, E. (1962) *Analytic Function Theory II*, Ginn and Company, Boston New-York.

Il'yashenko, Yu, S. & Yakovenko, S.Y. (1991) Smooth normal forms of local families of diffeomorphisms and vector fields, *Russian Math. Survey* **46**, pp. 1–43.

Kirrmann, P., Schneider, G. & Mielke, A. (1992) The validity of modulation equations for extended systems with cubic nonlinearities, *Proc. Math. Soc. Ed.* **122A**, pp. 85–91.

Maslov, V.P. (1976) *Operational Methods*. Mir, Moscow.

Mielke, A., & Schneider, G. (1995) Attractors for modulation equations on unbounded domains, existence and comparison, *Nonlinearity* **8**, pp. 743–768.

Protter, M. & Weinberger, H. (1967) *Maximum Principle in Differential Equations*. Prentice Hall, Englewood Cliffs N.J.

Stroock, D. (1988) Diffusion semigroups corresponding to uniformly elliptic divergence form operators. In Azéma, J., Meyer, P.A. & Yor, M. (eds.) *Séminaire de Probabilités XXII*, Lecture Notes in Mathematics 1321, Springer Verlag.

Takac, P., Bollerman, P., Doelman, A., van Harten, A. & Titi, E.S. (1996) Analyticity of essentially bounded solutions to semilinear parabolic systems and validity of the Ginzburg-Landau equation, *SIAM J. Math. Anal.* **27**, pp. 424–448.

van Harten, A. (1991) On the validity of Ginzburg-Landau's equation, *J. Nonlinear Science* **1**, pp. 397–422.

Wiener, N. (1967) *The Fourier Integral and Certain of its Applications*. Cambridge Univ. Press, Cambridge.

THREE LECTURES ON MATHEMATICAL FLUID MECHANICS

PETER CONSTANTIN
Department of Mathematics
The University of Chicago
Chicago, Illinois 60637
USA

1. Introduction

In these lectures I will describe some of the mathematical problems encountered in the study of incompressible fluid turbulence. The equations of motion, the Navier-Stokes equations,

$$(\partial_t + u \cdot \nabla - \nu\Delta)\, u + \nabla p = f$$

are an expression of Newton's law $F = ma$. The condition of incompressibility is

$$\nabla \cdot u = 0.$$

The fluid occupies a region in three dimensional or two dimensional Euclidean space. Position is denoted by $x \in \mathbf{R}^n$, $n = 2$ or 3, and time by t. The velocity $u(x,t)$ is a vector valued function, with values in the same Euclidean space $\mathbf{R}^n$. The pressure $p(x,t)$ is a scalar; its mathematical role is to maintain the incompressibility condition. The coefficient ν is a non-negative number, the kinematic viscosity. If $\nu = 0$ then the equations are called the Euler equations. The vector valued function f represents body forces. The equations are usually supplemented by initial conditions and by boundary conditions.

The questions of existence of solutions, their smoothness and time evolution are both natural and relevant. In addition to these questions, there exist questions that are more specific to turbulence, such as questions regarding the large Reynolds number asymptotics and the scaling exponents of various quantities computed from the basic fluid variables.

J. C. Robinson and P. A. Glendinning (eds.),
From Finite to Infinite Dimensional Dynamical Systems, 145–175.

The plan of the lectures is as follows. I will start with the question of existence of smooth solutions to the three dimensional incompressible Euler equations. The basic issue is whether finite time singularities arise from smooth and compactly supported initial data. This question is open. I will present some criteria for regularity. These provide necessary and sufficient conditions for blow up, conditions that are both testable numerically and relevant physically. In addition I will describe a class of models, which I called "active scalars" (as opposed to passive scalars, a more familiar object in fluid dynamics), that are useful because they can be formulated in two space dimensions, and hence are more easily visualised and more amenable to direct computer simulation. These models do not have to be restricted to two spatial dimensions; they have a common mathematical structure and cover a large part of the physically relevant incompressible fluid mechanics.

In the second lecture I will address the question of the inviscid limit. This is one of the most important problems in turbulence, namely to describe effective equations valid at high Reynolds numbers. As is the case with the global existence of smooth solutions, this also is an open problem. Well known difficulties are caused by so-called boundary layers, small regions in which there are sharp changes in the qualitative nature of solutions. As their name indicates, these layers originate at boundaries, places where the fluid is in contact with a solid or with a different fluid. If one considers an idealized situation in which there are no such boundaries (periodic boundary conditions, for instance), then the limiting behavior for a fixed period of time is well understood, *if the solutions are smooth.* I will address the question of the inviscid limit for solutions that are not smooth. In that case transition layers appear in which the smoothing effect of the viscosity is important and differences between the solutions of the Euler and Navier-Stokes equations are significant.

In the last lecture I will address the question of the temporal limit, that is, the long time behavior of solutions of Navier-Stokes equations, when the control parameters (viscosity, driving forces, boundary conditions) are kept fixed. The Reynolds number can be defined in terms of these control parameters, and is fixed. The initial data are not restricted in size, they are only required to belong to appropriate function spaces. Long time refers in this case to time scales that are large compared with all time scales, including viscous. The steady supply of energy to the system can result in complicated dynamics. Even for the case of steadily forced, two dimensional periodic Navier-Stokes equations, the question whether the long time dynamics is described by ordinary differential equations is still open. I will present two results. One concerns a class of initial data that play a role for the Navier Stokes equation that is similar to the one played by the trigonometric polynomials for the heat equation. The other result concerns the exponential decay of the temporal power spectrum.

2. First lecture: Regularity criteria for Euler equations and active scalars

When one sets the viscosity to zero in the Navier-Stokes equations one obtains the Euler equations. This is a conservative system, in the sense that the kinetic energy

$$\int |u(x,t)|^2\, dx$$

is constant in time, if the solution is smooth enough. (The minimal regularity needed for the conservation of kinetic energy is the object of a conjecture of Onsager; roughly speaking a Hölder exponent of 1/3 is sufficient (Constantin *et al.*, 1994a). In this lecture we will discuss smooth solutions. If the space dimension is two then the evolution starting from smooth initial data is well defined and global: the solution remains smooth for all time. For three space dimensions the known results are only local in time. The difference between the two cases is due to the presence in two dimensions of additional integrals of motion concerning the evolution of the vorticity. The mathematical framework can be presented as follows:

The Euler equations are equivalent to the requirement that two first order differential operators Ω and D_t, commute

$$[D_t, \Omega] = 0.$$

Ω is associated to a divergence-free function $\omega = \omega(x,t) \in \mathbf{R}^n$, $\nabla \cdot \omega = 0$:

$$\Omega = \omega(x,t) \cdot \nabla.$$

The characteristics of this operator are called vortex lines. The operator D_t is also associated to a divergence-free function $u = u(x,t) \in \mathbf{R}^n$, $\nabla \cdot u = 0$:

$$D_t = \frac{\partial}{\partial t} + u \cdot \nabla.$$

The characteristics of this operator are called particle paths. The vanishing of the commutator can be expressed as an evolution equation for the coefficients ω:

$$\left(\frac{\partial}{\partial t} + u \cdot \nabla\right) \omega = \omega \cdot \nabla u.$$

The equation is nonlinear because u and ω are coupled:

$$u = \mathcal{K} * \omega$$

A difference between $n = 2$ and $n = 3$ lies in the nature of the coupling matrix $\mathcal{K}$. For $n = 2$

$$\mathcal{K}_{ij}^{(2E)}(y) = \delta_{ij} \log\left(|y|\right),$$

where δ_{ij} is the Kronecker delta. In three dimensions

$$\mathcal{K}_{ij}^{(3E)}(y) = \epsilon_{ijk}\frac{\hat{y}_k}{|y|^2}$$

where $\hat{y}$ is the unit vector $y/|y|$, ϵ_{ijk} is the signature of the permutation $(1,2,3) \mapsto (i,j,k)$ and repeated indices are summed. Note that the strength of the singularity of $\mathcal{K}^{(3E)}$ at the origin is of the order $2 = n-1$ whereas the strength of the singularity of $\mathcal{K}^{(2E)}$ is merely logarithmic. This is a crucial difference. It reflects the fact that in the case of $n = 3$ the divergence-free function ω is the vorticity, i.e. the curl of velocity, $\omega = \nabla \times u$, whereas in the case of $n = 2$ the vector ω defining Ω is not the two-dimensional vorticity θ (which is a scalar, $\theta = \partial u_2/\partial x_1 - \partial u_1/\partial x_2$), but rather its gradient rotated by 90 degrees, $\omega = \nabla^{\perp}\theta$.

In general, if $n = 2$, the equation $[D_t, \Omega] = 0$ is equivalent to the time evolution of an active scalar (Constantin, 1992). Active scalars are functions $\theta(x,t)$ that obey

$$(\partial_t + u \cdot \nabla)\,\theta = 0$$

where u is obtained from ω as above,

$$u = \mathcal{K} * \omega,$$

and θ is related to ω by

$$\omega = \nabla^{\perp}\theta.$$

The natural analogue of the 3 dimensional coupling in a two dimensional model is

$$\mathcal{K}_{ij}^{(QE)} = \delta_{ij}\frac{1}{|y|}.$$

This coupling has the correct singularity strength $1 = n - 1$. It turns out that this is a physically significant model: the scalar θ represents temperature in a quasi-geostrophic (Coriolis forces balance pressure gradients) approximation of atmospheric flow. This defines the quasi-geostrophic active scalar equation (QE). The active scalar models in two space dimensions are a convenient class of relatively simple and computable systems that include, besides QE, the incompressible hydrodynamical equations appearing in infinite Prandtl number convection, the common two dimensional Euler equations, vortex patches, etc. The different singularities of $\mathcal{K}$ at the origin in different models result in differences in the qualitative behavior of solutions.

We will discuss now briefly the relation between the 3D Euler equations and QE in what concerns possible finite time blow up. Blow up is the appearance in finite time of singularities in a solution whose initial vorticity is C_0^∞ (smooth functions with compact support). We will use the same notation for both the three dimensional Euler equations and QE; we will mention differences when necessary.

The integral lines of Ω are material, i.e., they are carried by the flow. In QE the vortex lines are iso-θ lines. Their length element is $|\omega|$. It is well-known (see Kato *et al.* (1984) for the 3D incompressible Euler equations, but the result holds for QE as well) that

$$\int_0^T \|\omega(\cdot,t)\|_{L^\infty(dx)} dt < \infty$$

is a necessary and sufficient condition (BKM) for regularity. (If the initial vorticity is C_0^∞ and BKM holds then the vorticity remains C_0^∞ on $[0,T]$.) Therefore blow up cannot occur without rapid stretching of the vorticity. One can prove also that blow up cannot occur without the development of large spatial gradients of ω (Constantin, 1994a & b). A natural way to quantify this is by associating to any Hölder seminorm of vorticity,

$$|\omega(\cdot,t)|_\mu = \sup_{x\neq y} \frac{|\omega(x)-\omega(y)|}{|x-y|^\mu}.$$

(with $0 < \mu \leq 1$), the unique quantity $\ell_\mu(t)$ with dimensions of length which one can form from it and the L^2 norm. The kinematic inequality

$$\|\omega(\cdot,t)\|_{L^\infty} \leq \|u(\cdot,t)\|_{L^2} \left(\ell_\mu(t)\right)^{-\frac{5}{2}}$$

follows in a straightforward manner from the fact that $\omega(\cdot,t)$ is divergence-free using the identity $\omega = P\omega$ where P is the Leray–Hodge projector on divergence-free vectors ((15) in Constantin, 1994a). Therefore BKM implies that the integral of the right hand side must diverge for blow-up to occur (Theorem 1 in Constantin, 1994a). In order to check BKM one needs to study $|\omega|$. This evolves according to the stretching equation:

$$D_t|\omega| \;-\; \alpha|\omega|.$$

The stretching rate alpha is given by

$$\alpha(x) \;=\; (\nabla u(x))\,\xi(x)\cdot\xi(x)$$

and the direction field ξ by

$$\xi(x) \;=\; \frac{\omega(x)}{|\omega(x)|}.$$

The region $\{x : |\omega(x)| > 0\}$ is material (carried by particle paths). Both α and ξ are defined in it. The stretching rate α has a remarkable integral representation (Constantin, 1992 & 1994a)

$$\alpha(x) = \text{P.V.} \int D(\hat{y}, \xi(x+y), \xi(x))\, |\omega(x+y)|\, \frac{dy}{|y|^n}.$$

Here n is the dimension of space (3 for Euler equations, 2 for QE) and

$$D^{(3E)}(\hat{y}, \xi(x+y), \xi(x)) = (\hat{y} \cdot \xi(x))\text{Det}(\hat{y}, \xi(x+y), \xi(x))$$

(where $\text{Det}(a, b, c)$ means the determinant of the matrix formed by the column vectors a, b, c), and

$$D^{(QE)}(\hat{y}, \xi(x+y), \xi(x)) = (\hat{y} \cdot \xi^{\perp}(x))(\xi(x+y) \cdot \xi^{\perp}(x)).$$

Note that the geometric factors D vanish not only in the spherical average, but also if the vectors $\xi(x+y)$ and $\xi(x)$ are parallel or anti-parallel. More precisely, if $\cos\phi = \xi(x) \cdot \xi(x+y)$ then $|D| \leq |\sin\phi|$. The geometric factors vanish also if $\hat{y}$ is parallel to $\xi(x)$.

Based on these properties one can prove that if the direction field ξ is smooth in regions of high vorticity then blow up does not occur. This result generalizes the regularity result for two dimensional Euler equations because a locally planar flow has ξ constant in space, perpendicular to the plane of motion. The result relating smoothness of ξ to absence of blow up for the QE can be found in (Constantin *et al.*, 1994b), and for the Euler equations in (Constantin *et al.*, 1996). We consider a QE active scalar $\theta(x,t)$ with smooth (C_0^{∞}) initial datum and suppose that the scalar is defined and smooth for $(x,t) \in \mathbf{R}^2 \times [0,T)$. In the case of the Euler equations we consider a solution of the three dimensional incompressible Euler equations with smooth and compactly supported initial vorticity and assume that the solution is smooth on a time interval $0 \leq t < T$. Consider the particle trajectories $X(q,t)$, solutions of

$$\frac{dX}{dt} = u(X,t),$$

$$X(q,0) = q.$$

If W_0 is a set we denote by W_t its image at time t

$$W_t = X(W_0, t)$$

and by $\mathcal{O}_T(W_0)$ the semi-orbit:

$$\mathcal{O}_T(W_0) = \{(x,t); x \in W_t, 0 \leq t \leq T\}.$$

In the case of QE we say that a set W_0 is *smoothly directed* if there exists $\rho > 0$ such that the following two conditions are satisfied. First, for every $q \in W_0^*$,

$$W_0^* = \{q \in W_0; |\omega_0(q)| \neq 0\}.$$

and all $t \in [0,T)$, the function $\xi(\cdot,t)$ has a Lipschitz extension (denoted by the same letter) to the Euclidean ball of radius 4ρ centered at $X(q,t)$ and

$$M = \lim_{t\to T} \sup_{q\in W_0^*} \int_0^t \|\nabla\xi(\cdot,s)\|^2_{L^\infty(B_{4\rho}(X(q,s)))} ds < \infty.$$

And secondly,

$$\lim_{t\to T} \sup_{q\in W_0} \int_0^t |u(X(q,s),s)|^2 ds < \infty.$$

The result of (Constantin *et al.*, 1994b) for QE is

Theorem 1 *Assume W_0 is smoothly directed. Then*

$$\sup_{\mathcal{O}_T(W_0)} |\omega(x,t)| < \infty.$$

An example of a blow up scenario ruled out by this theorem is the appearance of a shock in θ over a smooth shock front. The modifications needed for the Euler equations are minor: We say that a set W_0 is *smoothly directed* for the Euler equations if there exists $\rho > 0$ and r, $0 < r \leq \frac{\rho}{2}$ such that the following three conditions are satisfied. First, for every $q \in W_0^*$,

$$W_0^* = \{q \in W_0; |\omega_0(q)| \neq 0\}.$$

and all $t \in [0,T)$, the function $\xi(\cdot,t)$ has a Lipschitz extension (denoted by the same letter) to the Euclidean ball of radius 4ρ centered at $X(q,t)$ and

$$M = \lim_{t\to T} \sup_{q\in W_0^*} \int_0^t \|\nabla\xi(\cdot,s)\|^2_{L^\infty(B_{4\rho}(X(q,s)))} ds < \infty.$$

Secondly,

$$\sup_{B_{3r}(W_t)} |\omega(x,t)| \leq m \sup_{B_r(W_t)} |\omega(x,t)|$$

holds for all $t \in [0,T)$ with $m \geq 0$ constant. And finally,

$$\sup_{B_{4\rho}(W_t)} |u(x,t)| \leq U$$

holds for all $t \in [0,T)$.

The first assumption means that the direction of vorticity is well behaved in a fixed neighborhood of a bunch of trajectories. The second assumption states that this neighborhood is large enough to capture the local intensification of ω. In other words, the competing and significantly stronger blow up trajectories, if they exist, do not come near this bunch. The result of (Constantin *et al.*, 1996) for the Euler equations is

Theorem 2 *Assume W_0 is smoothly directed. Then there exists $\tau > 0$ and Γ such that*

$$\sup_{B_r(W_t)} |\omega(x,t)| \leq \Gamma \sup_{B_\rho(W_{t_0})} |\omega(x,t_0)|$$

holds for any $0 \leq t_0 < T$ and $0 \leq t - t_0 \leq \tau$.

The case of osculating anti-parallel vortex lines is covered by the next result, valid both for the three dimensional Euler equations (Constantin *et al.*, 1996) and QE (Constantin *et al.*, 1994b). We say that the set W_0 is *regularly directed* if there exists $\rho > 0$ such that

$$\lim_{t\to T} \sup_{q\in W_0^*} \int_0^t K_\rho(X(q,s))ds < \infty$$

where

$$K_\rho(x) = \int_{|y|\leq\rho} |D(\hat{y},\xi(x+y),\xi(x))||\omega(x+y)|\frac{dy}{|y|^2}.$$

Theorem 3 *Assume that W_0 is regularly directed. Then*

$$\sup_{\mathcal{O}_T(W_0)} |\omega(x,t)| < \infty.$$

An example of a blow up scenario ruled out by this theorem is the collision of two antiparallel vortex lines. The numerical evidence of (Constantin *et al.*, 1994b) for QE strongly supports two statements: first that sharp fronts do form in finite time (sharp means large gradients of θ, i.e. large $|\omega|$; based on the present computations, blow up cannot be predicted). Secondly that there is a marked difference in the rate of development of these sharp fronts caused by the nature of ξ. There exist initial data for which ξ develops only antiparallel, regularly directed singularities. The formation of fronts is then depleted. For other initial data, a saddle point in θ provides a Lipschitz singularity in the ξ direction field. This is the source of much more intense growth of magnitude of omega.

I will mention here briefly the corresponding results in the presence of viscosity. In the three dimensional incompressible Navier-Stokes equations all regular solutions and even certain suitable weak solutions (suitable weak solution is here a technical term, see Caffarelli *et al.*, 1982) satisfy

$$\langle|\omega(x,t)||\nabla\xi(x,t)|^2\rangle \leq \frac{\epsilon}{\nu^2}$$

where $\nu > 0$ is the viscosity, and $\langle \cdots \rangle$ is space and time average (Constantin, 1990). The number ϵ is bounded above a priori in terms of the initial data and represents the average energy dissipation. This inequality implies that there exists a viscous mechanism that prevents exceedingly large $|\nabla\xi|$ in regions of high vorticity. This is, I believe, the mechanism responsible for the phenomenon of vortex reconnection. A consequence of the bound above is that regions of non-zero vorticity have square integrable $\nabla\xi$. But if one assumes a "little" more, namely that $\nabla\xi$ is locally bounded in regions of high vorticity then one can prove (Constantin & Fefferman, 1993) that the solution is smooth. More precisely, denote by $P^{\perp}_{\xi(x)}\xi(x+y)$ the projection of $\xi(x+y)$ on the plane perpendicular to $\xi(x)$ and assume
Assumption (A): There exist constants $\Omega > 0$ and $\rho > 0$ such that

$$|P^{\perp}_{\xi(x,t)}(\xi(x+y,t))| \leq \frac{|y|}{\rho}$$

holds if both $|\omega(x,t)| > \Omega$ and $|\omega(x+y,t)| > \Omega$, and $0 \leq t \leq T$.

Theorem 4 *Under the assumption (A) the solution of the initial value problem for the Navier–Stokes equation is smooth (C^∞) on the time interval* $[0,T]$.

The interpretation of these results is that vortex reconnection is a viscous mechanism by which blow up is averted.

3. Regularity criteria: sketch of one proof

We will consider the three dimensional incompressible Euler equations. The vorticity magnitude obeys

$$D_t|\omega| = \alpha|\omega|$$

where

$$D_t = \partial_t + u \cdot \nabla$$

is the material derivative, and the stretching rate α is given by a singular integral

$$\alpha(x) = \text{P.V.} \int D(\hat{y}, \xi(x+y), \xi(x))\, |\omega(x+y)| \frac{dy}{|y|^3}.$$

The geometric factor D is given by

$$D(\hat{y}, \xi(x+y), \xi(x)) = (\hat{y} \cdot \xi(x))\text{Det}(\hat{y}, \xi(x+y), \xi(x)).$$

Let us consider now a number $\rho > 0$ and decompose

$$\alpha(x) = \alpha_{\text{in}}(x) + \alpha_{\text{out}}(x)$$

where

$$\alpha_{\text{in}}(x) = \text{P.V.} \int \chi\left(\frac{|y|}{\rho}\right) \cdots$$

and

$$\alpha_{\text{out}}(x) = \text{P.V.} \int \left(1 - \chi\left(\frac{|y|}{\rho}\right)\right) \cdots$$

with $\chi(r)$ a smooth non-negative function of one positive variable satisfying $\chi(r) = 1$ for $0 \le r \le \frac{1}{2}$, $\chi(r) = 0$ for $r \ge 1$. It is easy to prove the estimate:

$$|\alpha_{\text{out}}(x)| \le C\rho^{-\frac{5}{2}} \|u_0\|_{L^2}.$$

Indeed, integrating by parts one obtains

$$|\alpha_{\text{out}}(x)| \le C\rho^{-1} \int_{|y| \ge \frac{1}{2}\rho} |u(x+y)| \frac{dy}{|y|^3}.$$

and the estimate follows from conservation of kinetic energy.

We are going to consider points x where ω does not vanish. Although regions of zero vorticity can be near regions of intense vorticity, no non-zero vorticity can appear during the Lagrangian evolution of a point of zero vorticity. We will assume that the direction field ξ has a Lipschitz extension (denoted still ξ) to an Euclidean ball of radius ρ. This ball might contain regions where ω vanishes.

Let us denote by $G = G_\rho(x)$ the maximum of the gradient of (the extension of) ξ in the ball of radius ρ centered at x:

$$G_\rho(x) = \sup_{|y| \le \rho} |\nabla \xi(x+y)|.$$

Clearly

$$D(\hat{y}, \xi(x+y), \xi(x)) \le G|y|$$

for $|y| \le \rho$. It follows that

$$|\alpha_{\text{in}}(x)| \le GI(x)$$

where

$$I(x) = \int \chi\left(\frac{|y|}{\rho}\right) |\omega(x+y)| \frac{dy}{|y|^2}.$$

Let us split the integral I again, using a very small $\delta \le \frac{\rho}{2}$:

$$I(x) = I_1 + I_2$$

where

$$I_1(x) = \int \chi\left(\frac{|y|}{\delta}\right) |\omega(x+y)| \frac{dy}{|y|^2}$$

and

$$I_2 = \int \left[1 - \chi\left(\frac{|y|}{\delta}\right)\right] \chi\left(\frac{|y|}{\rho}\right) |\omega(x+y)| \frac{dy}{|y|^2}.$$

We bound I_1 by

$$|I_1(x)| \le C\delta\Omega_\delta$$

where we set

$$\Omega = \Omega_\delta(x) = \sup_{|y|\le\delta} |\omega(x+y)|.$$

For the estimate of I_2 we use the fact that

$$|\omega| = \xi \cdot \omega$$

and the fact that $\omega = \nabla \times u$ and integrate by parts:

$$I_2(x) = \int u(x+y) \cdot \left\{\nabla \times \left[\xi \frac{1}{|y|^2} \chi\left(\frac{|y|}{\rho}\right) \left(1 - \chi\left(\frac{|y|}{\delta}\right)\right)\right]\right\} dy$$

Carrying out the differentiation we obtain four terms:

$$|I_2(x)| \le (A + B + D + E)$$

where

$$A \le CG \int_{|y|\le\rho} |u(x+y)| \frac{dy}{|y|^2}$$

$$B \le \frac{C}{\rho} \int_{|y|\le\rho} |u(x+y)| \frac{dy}{|y|^2}$$

$$D \le C \int |u(x+y)| \left[1 - \chi\left(\frac{|y|}{\delta}\right)\right] \chi\left(\frac{|y|}{\rho}\right) \frac{dy}{|y|^3}.$$

and

$$E \le \frac{C}{\delta} \int_{\frac{\delta}{2}\le|y|\le\delta} |u(x+y)| \frac{dy}{|y|^2}$$

Denote by U

$$U = U_\rho(x) = \sup_{|y|<\rho} |u(x+y)|.$$

Then it follows that

$$A \le CG\rho U,$$

$$B \le CU,$$

$$E \le CU$$

and

$$D \le CU \log\left(\frac{\rho}{\delta}\right).$$

Putting together these estimates we get

Lemma 1 *The logarithmic material rate of change of* $|\omega|$ *obeys*

$$|\alpha(x)| \leq A_\rho(x)\left[1 + \log\left(\frac{\rho}{\delta}\right)\right] + G_\rho(x)\left[\delta\Omega_\delta(x)\right]$$

where

$$A_\rho(x) = C\rho^{-\frac{5}{2}}\|u_0\|_{L^2} + CG_\rho(x)U_\rho(x)\left(1 + \rho G_\rho(x)\right)$$

and G_ρ, U_ρ, Ω_δ *are defined above.*

Now let us consider a bunch of trajectories $X(q,t)$ indexed by the set of initial points W_0: $q \in W_0$. (This set could be comprised of one single element, i.e., we could be speaking of one trajectory.) We assume that the solutions are smooth on the time interval $[0,T)$. We consider a neighborhood $\mathcal{B}_{4\rho}$ of the bunch of trajectories starting from W_0:

$$\mathcal{B}_{4\rho} = \{(x,t) :\ 0 \leq t < T,\ \exists\, q \in W_0,\ |X(q,t) - x| \leq 4\rho\}$$

We assume that $\mathcal{B}_{4\rho}$ is good in the sense that

$$\sup_{(x,t)\in\mathcal{B}_{4\rho}} |u(x,t)| \leq U < \infty$$

and

$$M = \lim_{t\to T}\sup_{q\in W_0^*}\int_0^t G_{4\rho}^2(X(q,s))ds < \infty.$$

Now let us recall the notation W_t for the slice of the bunch of trajectories at time t and let us denote by $B_r(W_t)$ an r - neighborhood ($2r \leq \rho$) of it:

$$B_r(W_t) = \{x :\ \exists\, q \in W_0,\ |x - X(q,t)| \leq r\}$$

Let us consider the time τ defined by

$$\tau = \frac{r}{4U}.$$

Denote by $w_r(t)$

$$w_r(t) = \sup_{B_r(W_t)} |\omega(x,t)|.$$

We assume that

$$w_{3r}(t) \leq m w_r(t)$$

Consider now $x \in B_r(W_t)$, $t < T$. The Lagrangian trajectory passing through x at time t is denoted $X(q',t)$. Note that q' might fail to belong to W_0. Nevertheless, if $r \leq \frac{\rho}{2}$ and $0 \leq t - s \leq \tau$ then $X(q',s) \in B_{2r}(W_s)$,

$$|X(q,s) - X(q',s)| \leq 2r \leq \rho$$

where $q \in W_0$. Then it follows that

$$G_\rho(X(q',s)) \le G_{4\rho}(X(q,s))$$

and, from Lemma 1 it follows that

$$|\alpha(X(q',s))| \le A_{4\rho}(X(q,s))\left[1+\log\left(\frac{\rho}{\delta}\right)\right] + G_{4\rho}(X(q,s))\left[\delta\Omega_\delta(X(q',s))\right]$$

Let us denote

$$\mathcal{A}(s) = \sup_{q\in W_0^*} A_{4\rho}\left(X(q,s)\right),$$

and

$$\mathcal{G}(s) = \sup_{q\in W_0^*} G_{4\rho}\left(X(q,s)\right).$$

it follows that

$$|\omega(X(q',t))| \le K\exp\int_{t_0}^{t}\left\{\mathcal{A}(s)\left[1+\log\left(\frac{\rho}{\delta}\right)\right] + \mathcal{G}(s)\left[\delta\Omega_\delta(X(q',s))\right]\right\}ds.$$

where

$$K = w_\rho(t_0),$$

$0 \le t - t_0 \le \tau$ and $\delta \le \frac{\rho}{2}$ is free, in particular it can be chosen to be less or equal to r:

$$\delta \le r.$$

Then in view of the fact that $X(q',s) \subset B_{2r}(W_s)$ and our assumptions it follows that

$$\Omega_\delta(X(q',s)) \le m w_r(s).$$

It follows that

$$w_r(t) \le K\exp\int_{t_0}^{t}\left\{\mathcal{A}(s)\left[1+\log\left(\frac{\rho}{\delta}\right)\right] + m\delta\mathcal{G}(s)w_r(s)\right\}ds$$

holds for any $0 < \delta \le r$, $0 \le t - t_0 \le \tau$. Let us define now the non-dimensional quantities

$$A = A(t,t_0) = \int_{t_0}^{t}\mathcal{A}(s)ds$$

and

$$Q = K\rho\int_0^T \mathcal{G}(s)ds.$$

Then consider

$$y(t) = \max_{t_0\le s\le t}\left(\frac{w_r(s)}{K}\right).$$

Let us choose

$$\frac{\rho}{\delta} = \max\left\{my(t)Q, \frac{\rho}{r}\right\}$$

It follows that

$$y(t) \le \left(\frac{\rho}{\delta}\right)^A \exp(1 + A)$$

Let us restrict τ further by demanding that

$$A(t, t_0) \le \frac{1}{2}$$

for $t - t_0 \le \tau$. Fix τ. It follows that

$$y(t) \le \max\{me^3 Q; \frac{\rho}{mrQ}\} = \Gamma$$

We proved

$$\sup_{B_r(W_t)} |\omega(x,t)| \le \Gamma \sup_{B_\rho(W_{t_0})} |\omega(x, t_0)|$$

holds for $0 \le t - t_0 \le \tau$, $t < T$, as stated in the corresponding theorem.

4. Second lecture: Finite time, small viscosity

In this lecture I will discuss the inviscid limit for non-smooth solutions. The need to study weak solutions arises when singularities play an important role. This can be the case if individual solutions with smooth initial data blow up in finite time (i.e. develop a finite time singularity); or, it can be the case that the long time behavior is best described by singular solutions. In this lecture I will describe results for the two dimensional Euler equations. These are equivalent to the active scalar equation

$$\left(\frac{\partial}{\partial t} + u \cdot \nabla\right)\theta = 0$$

where

$$u = \mathcal{K}^{(2E)} * \omega, \quad \omega = \nabla^{\perp}\theta.$$

Even if $\theta \in C_0^\infty$, the velocity is not square integrable in general. The kinetic energy is defined in this case by

$$E_\theta = -\int_{\mathbf{R}^2} \theta(x)\psi_\theta(x)dx$$

where

$$\psi_\theta(x) = \frac{1}{2\pi}\int_{\mathbf{R}^2} \log(|x - y|)\,\theta(y)dy$$

is the stream function. Thus $u = u_\theta$ where

$$u_\theta = \nabla^\perp \psi_\theta.$$

If the initial vorticity

$$\theta(x,0) = a(x)$$

is a smooth function with compact support then the solution of the two dimensional Euler exists for all time, is smooth, compactly supported and satisfies

$$E_{\theta(\cdot,t)} = E_a,$$

and

$$\int_{\mathbf{R}^2} F(\theta(x,t))dx = \int_{\mathbf{R}^2} F(a(x))dx$$

for all time t and all locally Lipschitz continuous functions F that vanish at the origin. Consider the interval I_θ

$$I_\theta = [\min \theta, \max \theta].$$

One has

$$I_{\theta(\cdot,t)} = I_a$$

and, defining the distribution of θ to be the Radon measure (see Evans & Gariepy, 1992, page 5) π_θ on $\mathcal{C}(I_\theta)$ given by

$$\int F(\gamma)\pi_\theta(d\gamma) = \int_{\{x;\theta(x)\neq 0\}} F(\theta(x))dx$$

then instead of the infinitely many constraints given above one can write simply

$$E_{\theta(\cdot,t)} = E_a,$$

and

$$\pi_{\theta(\cdot,t)} = \pi_a.$$

These conservation laws may be used to introduce appropriate units of time and length:

$$T_\theta = \left(\|\theta\|_{L^\infty(dx)}\right)^{-1}$$

and

$$L_\theta = \left(\frac{\|\theta\|_{L^1(dx)}}{\|\theta\|_{L^\infty(dx)}}\right)^{\frac{1}{2}}$$

The corresponding velocity unit is denoted U_θ:

$$U_\theta = \frac{L_\theta}{T_\theta}.$$

It is easy to check directly from definitions that

$$\|u_\theta\|_{L^\infty(dx)} \leq U_\theta$$

holds. It is natural, both for mathematical reasons and for physical ones (Miller, 1990; Robert, 1991) to consider the evolution in the vorticity space $\mathbf{Y}$

$$\mathbf{Y} = L^1(\mathbf{R}^2) \cap L_c^\infty(\mathbf{R}^2)$$

of bounded functions with compact support; the norm is the sum of the L^1 and L^∞ norms. The weak solution

$$S^E(t)a = \theta(x,t)$$

of the Euler equation, corresponding to the initial datum $\theta(x,0) = a \in \mathbf{Y}$, exists for all $t \in R$ and is unique (Yudovich, 1963). The equation is solved in the sense of distributions. The products $\theta \otimes u_\theta$ (the nonlinear terms in the equation) are defined pointwise. The quantities L_a, T_a, U_a, E_a and the distribution π_a are conserved by the Eulerian flow, i.e.

$$C_{S^E(t)a} = C_a$$

if C_a stands for any of these quantities. The velocity

$$u^E(x,t) = u_{S^E(t)a}$$

satisfies

$$\|u^E(\cdot,t)\|_{L^\infty} \leq U_a$$

for all $t \in R$.

We denote by S the strain matrix – the symmetric part of the gradient of velocity:

$$S(x,t) = \frac{1}{2}\left((\nabla u) + (\nabla u)^*\right).$$

The quantity

$$\mathcal{A}(t) = \int_0^t \|S(\cdot,s)\|_{L^\infty} ds$$

plays an important role. It controls not only the growth of the Lipschitz norm of particle trajectories and of the L^p norms of gradients of vorticity but also the L^2 operator norm of the Gateaux derivative of the velocity solution map. The class of initial vorticities for which the quantity $\mathcal{A}(t)$ is finite for all time is therefore included in the class of initial vorticities for which the velocity map $u_a \mapsto u^E(\cdot,t)$ is continuous in L^2. The only class of non-smooth functions $a \in \mathbf{Y}$ that are known to have $\mathcal{A}(t)$ finite for all time is that of vortex patches with *smooth* boundaries (Bertozzi & Constantin,

1993) or slight generalizations thereof. Vortex patches are a much studied class of examples of $a \in \mathbf{Y}$. The initial vorticity $a(x)$ is a simple function

$$a = \sum_{j=1}^{N} a_0^{(j)} \chi_{D_j(0)}$$

where $a_0^{(j)}$ are real constants and $\chi_{D_j(0)}$ are the characteristic functions of bounded, simply connected domains in $\mathbf{R}^2$. If, in addition, the boundaries of the individual domains $D_j(0)$ are $C^{1,\mu}$ for some $\mu > 0$ then we speak of smooth vortex patches. In that case the weak solution

$$S^E(t)a = \theta(x,t)$$

is given by

$$\theta(x,t) = \sum_{j=1}^{N} a_0^{(j)} \chi_{D_j(t)}$$

where the domains $D_j(t)$ have $C^{1,\mu}$ boundaries for all time. If the boundaries of the patches have initially some degree of smoothness, but not $C^{1,\mu}$, then the evolution is still well defined as above but the qualitative description of the boundary might degenerate in time. This has important consequences for the inviscid limit.

I will present now recent results (Constantin & Wu, 1995 & 1996) regarding the inviscid limit in the case of non-smooth two dimensional vorticities. The L^2 limit of solutions $u^{(NS)}$ of the Navier-Stokes equations in the case of vortex patch initial data was discussed in (Constantin & Wu, 1995). It was proved that, if the initial vorticity is a vortex patch with smooth boundary, then the difference $u^{(NS)} - u^{(E)}$ between the Navier-Stokes and Euler velocities corresponding to this initial datum is in L^2 and converges to zero at a rate proportional to $\sqrt{\nu}$. This is a slower rate of convergence than the rate ($O(\nu)$) of the inviscid limit for smooth solutions (McGrath, 1968; Kato, 1972; Beale & Majda, 1981; Constantin, 1986). The fact that there is a drop in the rate of convergence when one passes from the smooth to the non-smooth regime is not an artifact; there are elementary examples providing lower bounds. The paper by Constantin & Wu (1996) considers the L^p inviscid limit for vorticities. The motivation is given by the statistical equilibrium theory of vortices (Miller, 1990; Robert, 1991). We are mostly interested in long time, uniform bounds, i.e., bounds that are valid for many turnover times and that have an explicit rate of vanishing. The smoothing effect that is present in the Navier-Stokes equations is absent in the Euler equations. Because of this, internal transition layers prevent *uniform* L^p bounds for the difference between vorticities of solutions with the same non-smooth initial data. Therefore, it seems that a *pathwise* uniform Eulerian inviscid limit in this phase space is not possible. The term pathwise refers here to the comparison of individual solutions, paths that start from the same initial data. It turns out that in order to obtain uniform bounds one needs to consider non-pathwise bounds: the most convenient close companion to a solution of the Navier-Stokes equation might be a mollified Euler solution. To be more precise, if $S^{NS}(t)b$ represents the solution (vorticity) of the Navier-Stokes equation with initial vorticity $b \in \mathbf{Y}$, if $S^E(t)b$ represents the solution of the Euler equation and if one denotes by $f_\delta = f * \phi_\delta$ the convolution with a mollifier ϕ_δ (a function of the form $\delta^{-2}\phi(x/\delta)$, where $\phi \geq 0$ is smooth, compactly supported and has integral one), then a pathwise estimate concerns the difference $S^{NS}(t)b - S^E(t)b$ and non pathwise estimates concern differences $S^{NS}(t)b - S^E(t)b_\delta$ and $S^{NS}(t)b - \left(S^E(t)b\right)_\delta$. The latter is better suited for long time estimates. While $S^E(t)b_\delta$ solves the Euler equations, $\left(S^E(t)b\right)_\delta$ solves suitably modified Euler equations. Recall that the Navier-Stokes equations and the Euler equations in $\mathbf{R}^2$ are

$$\frac{\partial u}{\partial t} + u \cdot \nabla u = -\nabla p + \nu \Delta u$$

$$\nabla \cdot u = 0$$

where $\nu > 0$ in the case of the Navier-Stokes equations, $\nu = 0$ in that of the Euler equation. The corresponding vorticity

$$\theta = \nabla^\perp \cdot u$$

satisfies

$$\frac{\partial \theta}{\partial t} + u \cdot \nabla \theta = \nu \Delta \theta$$

and u can be recovered from θ via

$$u = u_\theta = \frac{1}{2\pi}\left(\nabla^\perp \log(|\cdot|)\right) * \theta.$$

One can prove uniform L^p bounds that vanish as $\nu^{s/2p}$ for the difference between Navier-Stokes and modified Euler solutions corresponding to initial data in Besov spaces $b \in \mathbf{Y} \cap B_2^{s,\infty}(\mathbf{R}^2)$ (see for instance Stein (1970), page 170, who calls them Lipschitz spaces and denotes $B_p^{s,q}$ by $\Lambda_s^{p,q}$; he is, however, isolated in this regard). The optimal mollification is over a distance of order $\delta \sim \sqrt{\nu}$, a fact that is consistent with the estimate for the smallest length scales in two dimensional turbulence. The *non-pathwise uniform* results can be used to obtain *non-uniform pathwise* results (that is, pathwise results without rates of convergence). In particular one can prove the strong pathwise convergence in L^p, $1 < p < \infty$. We start by estimating the difference between velocities of solutions of the Navier-Stokes equations and Euler equations. Assume that $a \in \mathbf{Y}$ and $b \in \mathbf{Y}$ are initial vorticities for the Euler and respectively Navier-Stokes equation. All quantities can be non-dimensionalized using the units of length L_b and T_b associated to the Navier-Stokes initial data. In particular, the viscosity can be measured in the corresponding units of $L_b^2 T_b^{-1}$. One obtains (Constantin & Wu, 1995 & 1996)

Theorem 5 *Let $a \in \mathbf{Y}$ be the initial vorticity for a solution of the Euler equations and $b \in \mathbf{Y}$ the initial vorticity for a solution of the Navier-Stokes equations with kinematic viscosity ν. If the corresponding velocities, u_a and u_b are such that $u_b - u_a$ is square integrable then*

$$\|u^{NS}(\cdot,t) - u^E(\cdot,t)\|_{L^2(\mathbf{R}^2)} \leq \left(\|u_b - u_a\|_{L^2(\mathbf{R}^2)} + \|a\|_{L^2(\mathbf{R}^2)}\sqrt{\nu t}\right) \exp\left(\mathcal{A}(t)\right)$$

holds for all $t \geq 0$ with

$$\mathcal{A}(t) = \int_0^t \left\| \frac{1}{2}\left(\left(\nabla u^E\right) + \left(\nabla u^E\right)^*\right)\right\|_{L^\infty} ds.$$

For general a, $b \in \mathbf{Y}$, $u_a - u_b$ is not square integrable. However, quite obviously

Proposition 1 *Assume that $b \in \mathbf{Y}$ and that $a = b_\delta$ where*

$$b_\delta = b * \phi_\delta$$

with $\phi_\delta(x) = \delta^{-2}\phi(\frac{x}{\delta})$ a standard mollifier. Then

$$\|u_a - u_b\|_{L^2(\mathbf{R}^2)} \leq C\delta\|b\|_{L^2(\mathbf{R}^2)}.$$

Consequently, if $a = b_\delta$ and $b \in \mathbf{Y}$ it follows that

$$\|u^{NS}(\cdot,t) - u^E(\cdot,t)\|_{L^2(\mathbf{R}^2)} \leq C\|b\|_{L^2(\mathbf{R}^2)}\left[\delta + \sqrt{\nu t}\right] \exp\left(\mathcal{A}_\delta(t)\right).$$

where $\mathcal{A}_\delta$ is computed on the Euler solution $S^E(t)b_\delta$.

We will keep the notation b for the initial vorticity for the Navier-Stokes evolution and a for that of the Euler evolution. In order to obtain uniform vorticity estimates one needs a certain degree of regularity of b:

$$b \in \mathbf{Y} \cap \left(\cup_{0<s<1} B_2^{s,\infty}(\mathbf{R}^2)\right)$$

where $B_2^{s,\infty}(\mathbf{R}^2)$ is the Besov space (see for instance, Peetre, 1986) of vorticities in $L^2(\mathbf{R}^2)$ that have an L^2 modulus of continuity that is Hölder continuous of order s. If $b \in B_2^{s,\infty}(\mathbf{R}^2)$ then

$$\delta\|\nabla b_\delta\|_{L^2(\mathbf{R}^2)} \leq C\left(\frac{\delta}{\rho_b}\right)^s \|b\|_{L^2(\mathbf{R}^2)}$$

where

$$\rho_b = \left(\frac{\|b\|_{L^2(\mathbf{R}^2)}}{N_s(b)} \right)^{\frac{1}{s}}$$

and

$$N_s(b) = \sup_{y \in \mathbf{R}^2} \frac{\|b(\cdot + y) - b(\cdot)\|_{L^2(\mathbf{R}^2)}}{|y|^s}.$$

Note that, if b is a vortex patch initial datum and if the boundary of the patch is smooth then $b \in B_2^{1/2,\infty}$. More generally,

Lemma 2 *Let $b = \chi_D$ be the characteristic function of a bounded domain whose boundary has box-counting (fractal) dimension not larger than $d < 2$:*

$$d_F(\partial D) \leq d.$$

Then

$$b \in B_p^{2-d/p,\infty}(\mathbf{R}^2)$$

for $1 \leq p < \infty$.

One can prove the following short time result:

Theorem 6 *Assume that $b \in \mathbf{Y} \cap B_2^{s,\infty}$ for some $0 < s < 1$. Consider*

$$\theta^{NS}(\cdot, t) = S^{NS}(t)b,$$

and

$$\theta^E(\cdot, t) = S^E(t)b_\delta$$

with $\delta = \sqrt{\nu T_b}$.

For every $\epsilon > 0$, there exists an absolute constant γ depending only on s and ϵ such that, if

$$0 \leq \frac{t}{T_b} \leq \gamma$$

then, for every $p \geq 2$ there exists a constant K_b depending on p and b alone, such that

$$\|\theta^{NS}(\cdot, t) - \theta^E(\cdot, t)\|_{L^p(\mathbf{R}^2)} \leq K_b \nu^{(s-\epsilon)/2p}$$

holds for all ν small enough.

In order to obtain a long time result one needs to know that

$$\limsup_{\delta \to 0} \mathcal{A}_\delta(t) < \infty.$$

Recall that this quantity is computed by solving a family of Euler equations. The map $\delta \mapsto \mathcal{A}_\delta(t)$ is not known to be upper semicontinuous. In other words, even in the class of vortex patches with smooth boundaries, one cannot rule out the possibility that there exists b, a time t and a sequence $\delta \to 0$ such that $\mathcal{A}(t) < \infty$ for the solution starting from b but $\lim_{\delta\to 0} \mathcal{A}_\delta(t) = \infty$. If this does not happen then the result above holds without loss of ϵ (i.e. one can obtain the bound $\leq K_b \nu^{s/2p}$) and without restriction on time. Remarkably though, the global estimates can be obtained if one reverses the order of operations and, instead of mollifying the initial datum and then solving the Euler equations, one rather solves first the Euler equations and then mollifies. Consider thus $b \in \mathbf{Y} \cap B_2^{s,\infty}$ and assume that

$$\mathcal{A}(t) < \infty$$

for $0 \leq t \leq T$. In view of the results of Bertozzi & Constantin (1993) this is the case if b represents a vortex patch with smooth boundaries. Now consider

$$\theta^E(x,t) = S^E(t)b$$

and mollify it, i.e. consider the function

$$\theta_\delta(x,t) = \left(S^E(t)b\right) * \phi_\delta.$$

The equation obeyed by the mollified vorticity $\theta_\delta = \left(S^E(t)b\right)_\delta$ is

$$\left(\partial_t + u_\delta \cdot \nabla\right)\theta_\delta = \nabla \cdot \tau_\delta(u^E, \theta^E).$$

where

$$\tau_\delta(v,w) = (v - v_\delta)(w - w_\delta) - r_\delta(v,w)$$

with

$$r_\delta(v,w)(x) = \int \phi(y)\left(v(x-\delta y) - v(x)\right)\left(w(x-\delta y) - w(x)\right) dy.$$

The three dimensional analogues of these formulae were first used in a proof of the Onsager conjecture (Constantin et al., 1994a). One takes $\delta = \sqrt{\nu T_b}$ and compares θ_δ to $\theta^{NS}(x,t) = S^{NS}(t)b$.

Theorem 7 *Let $b \in \mathbf{Y} \cap B_2^{s,\infty} \cap B_4^{s/2,\infty}$ with $0 < s < 1$. Consider the solution*

$$\theta^E(\cdot,t) = S^E(t)b$$

and assume that

$$\mathcal{A}(T) = \int_0^T \left\| \frac{1}{2}\left((\nabla u^E) + (\nabla u^E)^*\right)\right\|_{L^\infty} dt < \infty$$

and that the solution satisfies

$$\theta^E \in L^2(0,T; B_4^{s/2,\infty} \cap B_2^{s,\infty}).$$

Consider $\delta = \sqrt{\nu T_b}$ *and set*

$$\theta_\delta = \theta^E * \phi_\delta$$

and

$$u_\delta = u^E * \phi_\delta$$

where ϕ_δ *is an appropriate mollifier. Consider the solution*

$$\theta^{NS}(\cdot,t) = S^{NS}(t)b.$$

Then there exists K *(depending on* b *only) such that*

$$\|u^{NS} - u_\delta\|_{L^2} \le K\sqrt{\nu}$$

holds for all $0 \le t \le T$ *and* ν *small. Moreover, for every* $p \ge 2$ *there exists a constant* K_p *(depending only on* p *and the solution* $S^E(t)b$ *for* $0 \le t \le T$*) such that*

$$\|\theta^{NS}(\cdot,t) - \theta_\delta(\cdot,t)\|_{L^p} \le K_p \nu^{s/2p}$$

holds for all $0 \le t \le T$*. In particular, if* b *is a vortex patch with smooth* $(C^{1,\mu})$ *boundaries then*

$$\|S^{NS}(t)b - \left(S^E(t)b\right)_{\sqrt{\nu T_b}}\|_{L^p} \le K_p \nu^{1/4p}$$

holds for all $0 \le t \le T$.

As a consequence of the *non-pathwise, uniform* results one can obtain *pathwise, non-uniform* results. One needs a theorem from Constantin & Wu (1995):

Theorem 8 *Assume that the initial vorticity* $b \in L^1 \cap L^\infty$ *has compact support, included in the disk*

$$\{x : |x| \le L\}.$$

Then

$$\|\theta^{NS}(\cdot,t)\|_{L^\infty(Q_t)} \le \|b\|_{L^\infty} e^{-U_b L/\nu}.$$

where

$$Q_t = \left\{x : |x| \ge C\left(L + \frac{\nu}{U_b} + U_b t\right)\right\}$$

and C is an absolute constant. Moreover, if one sets

$$\left[\left[\frac{x}{\delta}\right]\right] = \sqrt{1 + \frac{|x|^2}{\delta^2}},$$

then, for any $\delta > 0$

$$\|\theta^{NS}(\cdot, t)e^{[[\frac{\cdot}{\delta}]]}\|_{L^p} \le \|b(\cdot)e^{[[\frac{\cdot}{\delta}]]}\|_{L^p} e^{(U_b t/\delta)+(7\nu t/\delta^2)}$$

holds for any p, $1 \le p \le \infty$.

The pathwise results are:

Theorem 9 *Let* $b \in \mathbf{Y} \cap B_2^{s,\infty}$, $0 < s < 1$, *be an initial vorticity and consider* $\theta^{NS} = S^{NS}(t)b$. *There exists an absolute constant* γ *such that, for all* $t \in [0, \gamma T_b]$ *and all Lipschitz functions* f *that vanish at the origin*

$$\lim_{\nu \to 0} \int f(\theta^{NS}(x, t))\, dx = \int f(b(x))\, dx$$

holds. Consequently the weak limit of distributions is

$$\lim_{\nu \to 0} \pi_{[S^{NS}(t)b]}(dy) = \pi_b(dy).$$

The same result holds for arbitrary time interval $[0, T]$ *provided the solution of the Euler equation* $\theta^E = S^E(t)b$ *satisfies the assumptions*

$$b \in \mathbf{Y} \cap B_2^{s,\infty} \cap B_4^{\frac{s}{2},\infty},$$

$$\mathcal{A}(T) = \int_0^T \left\| \frac{1}{2}\left((\nabla u^E) + (\nabla u^E)^*\right) \right\|_{L^\infty} dt < \infty$$

and

$$\theta^E \in L^2(0, T; B_4^{s/2,\infty} \cap B_2^{s,\infty}).$$

As a consequence, the strong limit

$$\lim_{\nu \to 0} S^{NS}(t)b = S^E(t)b$$

holds in the L^p *norm for vorticities, for all* $1 < p < \infty$ *and the time intervals corresponding to the two situations considered above.*

Let us provide a sketch of the proof. In view of a result that was proved in Constantin & Wu (1995) we have

$$\int_{\mathbf{R}^2} f(\theta^{NS}(x, t))dx \simeq \int_{\mathbf{R}^2 \setminus Q_t} f(\theta^{NS}(x, t))dx.$$

(We use $\simeq$ to denote quantities that become equal in the limit $\nu \to 0$.) In view of the fact that

$$\int_{\mathbf{R}^2\setminus Q_t} f\left(\left(\theta^E(x,t)\right)_\delta\right) dx \simeq \int_{\mathbf{R}^2\setminus Q_t} f(\theta^E(x,t))dx = \int_{\mathbf{R}^2} f(\theta^E(x,t))dx$$

one needs to show that

$$\int_{\mathbf{R}^2\setminus Q_t} f\left(\left(\theta^E(x,t)\right)_\delta\right) dx \simeq \int_{\mathbf{R}^2\setminus Q_t} f(\theta^{NS}(x,t))dx.$$

Using the Lipschitz condition for f and the fact that $\mathbf{R}^2 \setminus Q_t$ is compact it is enough to have

$$\int_{\mathbf{R}^2\setminus Q_t} \left|\theta^{NS}(x,t) - \left(\theta^E(x,t)\right)_\delta\right|^2 dx \simeq 0.$$

But this follows from the non-pathwise estimates

$$\|\theta^{NS} - \theta_\delta\|_{L^2} \leq K\nu^{s/4}.$$

The proof of the L^p convergence follows from the fact that the norms converge ($f(y) = |y|^p$)

$$\lim_{\nu\to 0} \|S^{NS}(t)b\|_{L^p} = \|S^E(t)b\|_{L^p}$$

and the strong convergence of velocities in L^2 (Constantin & Wu, 1995). This last fact implies that any weak limit as $\nu \to 0$ of a sequence $S^{NS}(t)b$ for fixed b and t must be $S^E(t)b$. Strong convergence in L^p for $1 < p < \infty$ follows from weak convergence and convergence of the norms because these spaces are uniformly convex.

The moral of this story is that the inviscid limit is not trivial if one works in the space defined by the natural constraints of the Euler equation. Even if one restricts oneself to subspaces, the pathwise limit is not uniform and one has to consider non-pathwise limits. The rates of convergence in these limits depend on detailed information about the initial data, and are likely to degenerate in time.

5. Third lecture: Long time behavior

In this lecture I will consider the case of spatially periodic, steadily forced two dimensional Navier-Stokes equations. The viscosity, spatial period and driving force f are fixed, and the time is allowed to tend to infinity. The equations can be written (see, for instance, Constantin & Foias (1988) and references therein) in abstract form:

$$\frac{du}{dt} + Au + B(u,u) = f.$$

The function u belongs to H, a real Hilbert space

$$\left(L^2\right)^2 \cap \{u; \nabla \cdot u = 0\}$$

with scalar product $(\cdot,\cdot)_H$ and norm $|\cdot|_H$. The functions are periodic with period 1. The nature of the driving forces f is the only factor that determines the long time dynamics. The linear self-adjoint operator A is the Stokes operator. In this simple situation it coincides with the Laplacian acting on each component. We are considering functions of mean zero. The eigenvalues of A are denoted λ_j, $0 < \lambda_1 \leq \lambda_2 \cdots$. The inverse of A is compact, $\lambda_j \to \infty$. The nonlinearity is quadratic

$$B(u,u) = \mathbf{P}(u \cdot \nabla u)$$

where $\mathbf{P}$ is the Leray-Hodge projector on divergence free functions. The solution of the Navier-Stokes equations are as smooth as the forces f allow (think of the equation $Au = f$). The map that assigns to an initial velocity u_0 the solution at time t is denoted (in this section !) by

$$u_0 \mapsto S(t)u_0.$$

It is a nonlinear semigroup,

$$S(t) : H \to H \quad \text{continuous},$$

$$\lim_{t \to 0} S(t)u_0 = u_0$$

$$S(t+s)u_0 = S(t)S(s)u_0.$$

The semigroup has a compact, convex, absorbing set $\mathcal{B}$. Absorbing means: for any bounded set F of initial data there exists a finite time t_F such that, for all $t \geq t_F$, the image of the set under $S(t)$ is included in $\mathcal{B}$, $S(t)F \subset \mathcal{B}$. The omega limit set of the absorbing set is given by

$$\mathcal{A} = \cap_{t>0} S(t)\mathcal{B}.$$

This set is the global (universal) attractor. It is compact, connected, invariant

$$S(t)\mathcal{A} = \mathcal{A},$$

and contains all omega limit sets of all bounded sets: If F is bounded in H then

$$\omega(F) \subset \mathcal{A}$$

where

$$\omega(F) = \cap_{t>0} \mathrm{cl}_H \left(\cup_{s \geq t} S(s)F\right)$$

and cl_H means closure in H. In particular all trajectories $S(t)u_0$ converge to $\mathcal{A}$. Thus the temporal limit leads to $\mathcal{A}$. Moreover, all Borel measures in function space that are invariant under the semigroup are supported in $\mathcal{A}$. This set has finite fractal dimension (Constantin *et al.*, 1988), and the dimension can be understood as the number of degrees of freedom in the spirit of Landau (Landau & Lifschitz, 1959). The fact that the dimension of the global attractor is finite is not sufficient to conclude that the dynamics are truly those of a finite system of ordinary differential equations. To this date, this question remains open for the Navier-Stokes equation.

It is natural to try to understand the function $f \mapsto \mathcal{A}$. Consider the set of global solutions

$$\mathcal{G} = \{u_0 \in H : \ S(t)u_0 \text{ extends to } t \in \mathbf{R}\}.$$

The global attractor can be characterized as

$$\mathcal{A} = \{u_0 \in \mathcal{G} : \ S(t)u_0 \text{ is bounded for } t \in \mathbf{R}\}.$$

Unfortunately, it is not easy to determine which solutions will be extendable globally backward in time. The linear heat equation provides a simple example one can use as a guide. While $\mathcal{G}$ is a somewhat complicated affine space, the attractor for the linear heat equation $\frac{du}{dt} + Au = f$ is just $\mathcal{A} = \{A^{-1}f\}$. It is natural, in view of the above characterization of the attractor to consider the set of those global solutions that grow at most exponentially backward in time

$$\mathcal{M} = \left\{u_0 \in \mathcal{G} : \lim \sup_{t \to -\infty} \frac{\log |S(t)u_0|_H}{|t|} < \infty \right\}.$$

This set can be described in terms of the Dirichlet quotient

$$\lambda(u) = \frac{(Au, u)_H}{(u, u)_H}$$

as:

$$\mathcal{M} = \left\{u_0 \in \mathcal{G} : \lim \sup_{t \to -\infty} \ \lambda(S(t)u_0) < \infty \right\}.$$

For the linear heat equation the set $\mathcal{M}$ is the translation by $A^{-1}f$ of the set of trigonometric polynomials, a dense affine set. One can prove (Constantin *et al.*, 1997) that also the $\mathcal{M}$ corresponding to the Navier-Stokes semigroup is weakly dense in H. The Dirichlet quotient is associated to an average length. In the absence of forces, $f = 0$ this length increases in time (λ decreases):

$$\frac{d}{dt}\lambda + |(A - \lambda)v|_H^2 = 0.$$

As time tends to infinity, if $f = 0$, then the solution $u(t) = S(t)u_0$ will tend exponentially to zero. Its Dirichlet quotient $\lambda = \lambda(u(t))$ will tend to one of the eigenvalues λ_j of the Stokes operator. Moreover, the normalized function v appearing in the equation above,

$$v = \frac{u}{|u|_H}$$

will tend to one of the eigenfunctions w_j of the Stokes operator corresponding to the eigenvalue λ_j (Foias & Saut, 1984). The identity

$$AB(u,u) = B(u, Au) - B(Au, u)$$

implies that any eigenfunction of the Stokes operator is a time independent solution of the Euler equations. In the case $f \neq 0$ the forward evolution is toward a possibly complicated but finite dimensional attractor. The dynamics on $\mathcal{M}\setminus\mathcal{A}$ can be related to the inviscid Eulerian dynamics: rescaling appropriately velocity and time, one obtains the Euler equations as the infinite negative time limit. More precisely, one considers the change of time

$$\frac{dT}{dt} = \frac{|u(t)|_H}{|u_0|_H}$$

for $u_0 \in \mathcal{M} \setminus \mathcal{A}$. Then one considers the function

$$w(T) = v(t) = \frac{u(t)}{|u(t)|_H};$$

one can prove (Constantin *et al.*, 1997)

Theorem 10 *If $u_0 \in \mathcal{M} \setminus \mathcal{A}$ then from any sequence $T_j \to -\infty$ one can extract a subsequence (denoted for simplicity also T_j) such that the sequence of functions*

$$w_j(T) = w(T + T_j)$$

converges in $L^2_{loc}(\mathbf{R}, H)$ to a global solution of the Euler equations. In addition, for every $u_0 \in \mathcal{M} \setminus \mathcal{A}$ there exists a sequence $T_j \to -\infty$ such that the corresponding sequence $w_j(T)$ converges to a stationary solution of the Euler equations.

The simplest possible scenario is that for every $u_0 \in \mathcal{M}\setminus\mathcal{A}$ and any sequence $T_j \to -\infty$ the corresponding sequence $w_j(T)$ converges to an eigenfunction of the Stokes operator depending only on u_0, while λ converges to the corresponding eigenvalue. This has not been proved or disproved yet.

As the dimension of the attractor becomes large for large Reynolds numbers one might wonder if there are any predictions one can make based on the equations of motion. In this section we are talking about the temporal limit; such predictions will concern the so called dissipative range. One of the most common ways to analyze temporal data, such as the temperature $\theta(t)$ recorded at a given location in the Rayleigh-Bénard convection experiment is to compute the power spectrum,

$$P(\omega) = \lim_{T\to\infty} \frac{1}{T} \left| \int_0^T e^{-i\omega t}\theta(t)dt \right|^2 .$$

This is a much studied and used procedure; its mathematical justification is based on classical work on stochastic signals, going back to Khintchin, Kolmogorov and Wiener. At finite Rayleigh and Reynolds numbers though, the signal is not a random process, rather it is the output of a nonlinear partial differential equation. Using the properties of solutions of Navier-Stokes equations one can still make mathematical sense of the procedure. Moreover one can prove that the power spectrum must decay at least exponentially at high frequencies (Bercovici *et al.*, 1995). Specifically, consider the finite time auto-correlation function

$$C_T(t) = \frac{1}{T} \int_0^T \theta(t+s)\theta(s)ds$$

of a scalar quantity θ that might be the temperature at a fixed location or a certain velocity component at a fixed location. Consider also the finite time power spectrum,

$$P_T(\omega) = \frac{1}{T} \left| \int_0^T e^{-i\omega t}\theta(t)dt \right|^2 .$$

It is easy to check that, in the sense of distributions, $C = \lim_{T\to\infty} C_T$ exists if and only if $P = \lim_{T\to\infty} P_T$ exists, in which case $\hat{C} = P$ holds where $\hat{C}$ is the Fourier transform in the sense of distributions. Consider invariant Borel measures μ (Borel refers to the topology of H, invariant means $\mu(S(t)^{-1}B) = \mu(B)$ for all $t \geq 0$ and all Borelian subsets $B \subset H$). Such measures exist, for instance by considering the Bogoliubov-Krylov procedure (Krylov & Bogoliubov, 1937), adapted for infinite dimensions:

$$\int \Phi(u)\mu(du) = LIM_{T\to\infty} \frac{1}{T} \int_0^T \Phi(S(t)u_0)dt$$

defines an invariant measure μ, for each $u_0 \in \mathcal{D}(A)$ in the domain of the Stokes operator provided we understand LIM in the sense of the Banach limit: see Bercovici *et al.* (1995) for details, where it is proved that

Theorem 11 *There exists a positive constant $\tau = \tau(\mathcal{A})$ such that, for every $u_0 \in \mathcal{D}(A)$ and every choice of location x_0 and component i, the scalar*

$$\theta(t) = u_i(x_0, t)$$

has a power spectrum P which is a finite positive Borel measure such that

$$C(t) = LIM_{T\to\infty} C_T(t) = \frac{1}{2\pi}\int e^{i\omega t} P(d\omega)$$

holds with LIM in the sense of the Banach limit. Moreover

$$\int_{-\infty}^{\infty} e^{\tau|\omega|} P(d\omega) < \infty.$$

For any invariant Borel probability measure μ, the classical limit

$$C(t) = \lim_{T\to\infty} C_T(t)$$

exists $\mu(du_0)$- a.e.

The positive constant $\tau = \tau(\mathcal{A})$ depends only on the attractor, in other words only on the control parameters. The result holds also for the Boussinesq system in two dimensions, and in particular $\theta(t)$ can be the temperature recorded at some location. The same result holds in three dimensions if one assumes that the velocity is a bounded function. The result proves exponential decay of the power spectrum at high frequencies. Thus, slower than exponential decay – found by numerical treatment of experimental Rayleigh-Bénard convection data – can not persist at high temporal frequencies, for any fixed Rayleigh number.

6. Conclusions

In these lectures we have discussed three open problems. The first lecture addressed the possibility of finite time blow-up in the three dimensional Euler equations. We presented geometric necessary conditions for blow up, expressed in terms of the direction of vorticity.

The second lecture concerned the inviscid limit in the space defined by the natural constraints of the Euler equation. These are not spaces of smooth functions. The open problem here is whether the Euler equations can describe adequately this limit. We proved convergence results for solutions in appropriate Besov spaces. Even if one restricts oneself to such spaces, the pathwise limit is not uniform and one has to consider non-pathwise limits. The rates of convergence in these limits depend on detailed information about the initial data, and are likely to degenerate in time.

The third lecture addressed the problem of long time behavior of solutions of the two dimensional Navier-Stokes equations. The open problem here is whether this long time behavior can be adequately described by finitely many ordinary differential equations. We described a weakly dense set of initial data for which the Euler equations can describe the temporal behavior at large energies.

References

Beale, J. T. & Majda, A. (1981) Rates of convergence for viscous splitting of the Navier-Stokes equations, *Math. Comp.* **37**, pp. 243–259.

Bercovici, H., Constantin, P., Foias, C. & Manley, O. (1996) Exponential decay of the power spectrum of turbulence, *J. Stat. Phys.* **80**, pp. 579–602.

Bertozzi, A. & Constantin, P. (1993) Global regularity for vortex patches, *Comm. Math. Phys.* **152**, pp. 19–28.

Caffarelli, L., Kohn, R. & Nirenberg, L. (1982) Partial regularity of suitable weak solutions of the Navier-Stokes equations, *Commun. Pure and Appl. Math.* **35**, pp. 771–831.

Constantin, P. (1992) *Arg. Natl. Lab. preprint*, ANL/MCS-TM-170.

Constantin, P. (1994a) Geometric Statistics in Turbulence, *SIAM Review* **36**, pp. 73–98.

Constantin, P. (1994b) Geometric and analytic studies in turbulence, in *Lectures in Appl. Math.* **100**, Springer-Verlag.

Constantin, P. (1986) Note on loss of regularity for solutions of the 3D incompressible Euler and related equations, *Commun. Math. Phys.* **104**, pp. 311–326.

Constantin, P. (1990) Navier-Stokes equations and area of interfaces, *Commun. Math. Phys.* **129**, pp. 241–266.

Constantin, P., E, W., & Titi, E. (1994a) Onsager's conjecture on the energy conservation for solutions of Euler's equation, *Commun. Math. Phys.* **165**, pp. 207–209.

Constantin, P., Kukavica, I., Foias, C. & Majda, A. (1997) Dirichlet quotients and periodic 2D Navier-Stokes equations, *J. Math. Pures Appl.* **76**, pp. 125–153.

Constantin, P. & Fefferman, Ch. (1993) Direction of vorticity and the problem of global regularity for the Navier-Stokes equations, *Indiana Univ. Math. Journal* **42**, pp. 775.

Constantin, P., Fefferman, Ch. & Majda, A. (1996) Geometric constraints on potentially singular solutions for the 3D Euler equations, *Commun. PDE* **21**, pp. 559-571.

Constantin, P. & Foias, C. (1980) *Navier-Stokes Equations.* The University of Chicago Press, Chicago.

Constantin, P, Foias, C. & Temam, R. (1988) On the dimension of the attractors in two dimensional turbulence, *Physica D* **30**, pp. 294–296.

Constantin, P., Majda, A. & Tabak, E. (1994b) Formation of strong fronts in the 2-D quasigeostrophic thermal active scalar, *Nonlinearity* **7**, pp. 1495–1533.

Constantin, P. & Wu, J. (1995) Inviscid limit for vortex patches, *Nonlinearity* **8**, pp. 735–742.

Constantin, P. & Wu, J. (1996) The Inviscid limit for non-smooth vorticity, submitted.

Evans, L.C. & Gariepy, R.F. (1992) *Measure Theory and Fine Properties of Functions.* Studies in Advanced Mathematics, CRC Press, Boca Raton.

Foias, C. & Saut, J-C. (1984) Asymptotic behavior as time tends to infinity of solutions of the Navier-Stokes equations and nonlinear spectral manifolds, *Indiana University Math. Journal* **33**, pp. 459–477.

Kato, T. (1972) Nonstationary flows of viscous and ideal fluids in $\mathbf{R}^3$, *J. Funct. Anal.* **9**, pp. 296–305.

Kato, T., Beale, J.T. & Majda, A. (1984) Remarks on the breakdown of smooth solutions for the 3-D Euler equations, *Commun. Math. Phys.* **94**, pp. 61–66.

Krylov, N. & Bogoliubov, N.N. (1937). La théorie génerale de la mesure, *Ann. Math.* **38**, pp. 65–113.
Landau, L.D. & Lifschitz, E.M. (1959) *Fluid Mechanics*. Pergamon Press.
Miller, J. (1990) Statistical mechanics of Euler equations in two dimensions, *Phys. Rev. Lett.* **65**, pp. 2137–2140.
McGrath, F. (1968) Nonstationary plane flow of viscous and ideal fluids, *Arch. Rational Mech. Anal.* **27**, pp. 329–348.
Peetre, J. (1986) *New Thoughts on Besov Spaces*. Duke University, N.C.
Robert, R. (1991) A maximum-entropy principle for two dimensional perfect fluid dynamics, *J. Stat. Phys.* **65**, pp. 531–551.
Stein, E.M. (1970) *Singular Integrals and Differentiability Properties of Functions*. Princeton University Press.
Yudovich, V.I. (1963) Non-stationary flow of an ideal incompressible liquid, *Zh. Vych. Mat.* **3**, pp. 1032–1066 (in Russian).

LOW-DIMENSIONAL MODELS OF TURBULENCE

or, The Dynamics of Coherent Structures

P.J. HOLMES, J.C. MATTINGLY AND R.W. WITTENBERG
Program in Applied and Computational Mathematics
Fine Hall, Washington Road
Princeton, NJ 08544
U.S.A.

1. Introduction and Contents

In these lecture notes, we outline an approach to the construction of models of turbulent energy production. We focus on flows energetically dominated by coherent structures and seek low-dimensional dynamical systems describing the interactions among small sets of such structures, which we then study via techniques of dynamical systems theory, paying particular attention to symmetries inherited from the original physical system and governing evolution equations. Analyzing the bifurcations, invariant structures and attractors in the phase spaces of these (relatively) tractable systems, we hope to gain insight into physical mechanisms of turbulence production.

Before introducing our tools and techniques, we set the scene in section 2, with a brief discussion of coherent structures and the problem of turbulence, and section 3, with an overview of the dynamical systems viewpoint. A major tool in our analysis is the *Karhunen-Loève* or *proper orthogonal decomposition (POD)*, which provides a basis for representations of complex spatio-temporal fields that is optimal in the sense that it converges (in L^2-norm) faster on average than any other representation. We discuss this in section 4. We then select a subspace spanned by the dominant *empirical eigenfunctions* thus produced, project the governing equations onto it, model to account for neglected effects due to truncation or spatial localization, and thus derive a relatively small set of ODEs. This process is illustrated in sections 5 and 6 for the near wall region of the turbulent boundary layer, and a sketch of the behavior of some of the models so produced is given in section 7.

J. C. Robinson and P. A. Glendinning (eds.),
From Finite to Infinite Dimensional Dynamical Systems, 177–215.

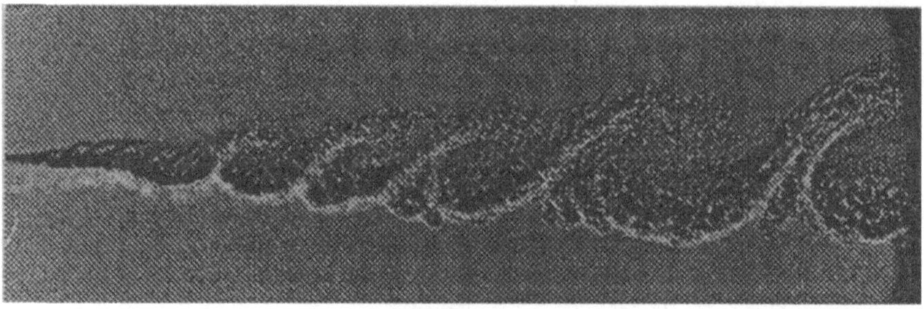

Figure 1. A shadowgraph image of coherent structures in the turbulent mixing layer between Helium at 1015 cm/s and Nitrogen at 384 cm/s, at a pressure of 8 atm. and Reynolds number $Re \sim 10^6$ (Brown & Roshko, 1974).

To appreciate and illustrate better the rôle of symmetry and one particular expression of it — heteroclinic cycles — that appears in our models, in sections 8–10 we discuss symmetries in general, and the group $O(2)$ in particular, and show how it appears in a simpler and more accessible nonlinear scalar PDE, the Kuramoto-Sivashinsky equation, which shares some of the same behavior as the boundary layer models. In section 11 we mention random perturbation of heteroclinic cycles, prompted by the fact that the effects of the outer boundary layer flow on the near wall region of our model may be replaced by a quasi-random pressure field at the outer edge of that region. Finally, section 12 contains a brief summary.

In the space available here we can merely indicate some of the key ingredients and ideas in an approach which involves several areas of mathematics and considerable knowledge of fluid mechanics and turbulence. For a fuller review and critical survey of this approach, we refer the interested reader to the book of Holmes, Lumley & Berkooz (1996) and the extensive literature cited therein, to which these notes are heavily indebted.

2. Turbulence and Coherent Structures

Over the last twenty or thirty years, numerous studies of turbulent fluid motion have revealed aspects of organized motion in a wide variety of flows; that is, there are large scale ordered structures which, although neither steady in space nor time, persistently appear, disappear and reappear (see the articles in Lumley (1990)- including Cantwell (1990) and Holmes (1990) - or Robinson (1991)). The "roll" structures observed by Brown & Roshko (1974) in the mixing layer provide a particularly clear example; see figure 1.

Flow visualization techniques and simulations have revealed *coherent structures* such as those in figure 1 quite clearly, but the structures typ-

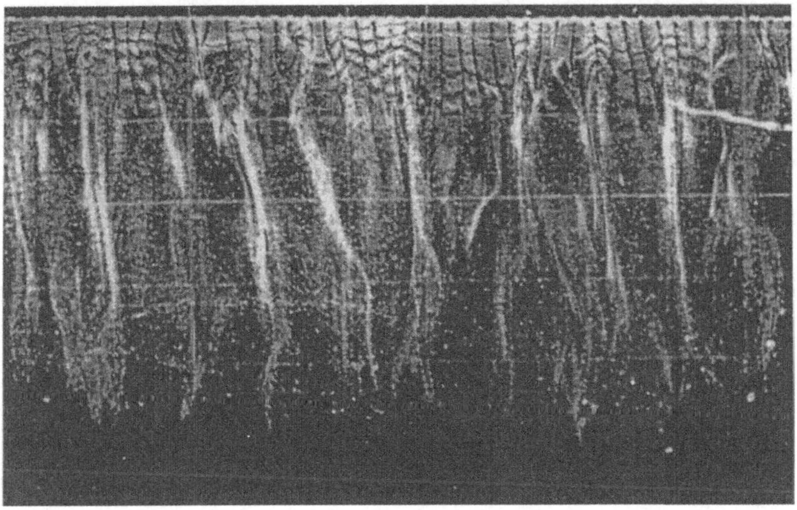

Figure 2. Streaks in the wall region of a turbulent boundary layer, extending in the direction of the mean flow. Hydrogen bubbles are released into the fluid at evenly spaced intervals from a thin wire and photographed from above, looking towards the wall (Kline *et al.*, 1967).

ically vary considerably in space and time. This has made them hard to pin down, so that sensitive conditional sampling techniques have been developed for their detection. Furthermore, their variability has precluded general agreement on their nature or even on a definition of coherent structures. However, it seems likely that such persistent macroscopic structure amidst all the small-scale activity and turbulent fluctuations provides a "backbone" for the flow, and there is the hope that *analysis of the dynamics of these structures may provide a basis for improved understanding of some aspects of turbulence*; it is this hope that motivated the study described in the present notes.

We shall be especially interested in the wall region of the turbulent boundary layer, where the presence of the wall leads to a structure which is strongly three-dimensional, highly intermittent and in general much more complex than in free shear flows. These structures involve persistent longitudinal (streamwise) vortices and low speed "streaks," visualised via hydrogen bubble traces in figure 2 (Kline *et al.*, 1967). For future reference, it is useful to establish the coordinates in the wall region flow domain — see figure 3 — in which we distinguish the homogeneous streamwise x_1, spanwise x_3, and inhomogeneous wall normal x_2 directions, with the mean flow U in the streamwise direction. The most intense motions are at a scale which is small compared to the depth of the boundary layer; this will provide the motivation later for artificially limiting the domain in the x_2 direction to the region very near the wall, up to $X_2 = 40$ in the nondimensional wall

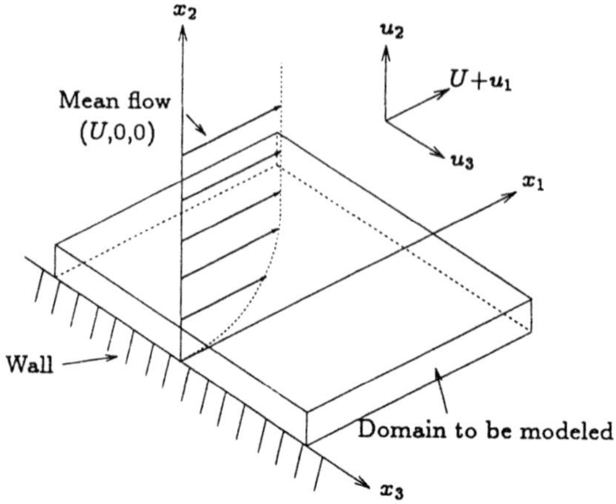

Figure 3. The wall region: Coordinates and modeling domain.

units appropriate to scaling of the inner wall layer (Tennekes & Lumley, 1972). The typical space-time evolution of the observed coherent structures is complicated, often exhibiting a repetitive cycle of events, including the lift, oscillation and ejection of longitudinal boundary layer streaks, followed by sweep and reformation. A major characteristic activity of interest is the burst/sweep cycle, in which a "burst" of low-(streamwise-)speed fluid being forced up away from the wall ($u_1<0$, $u_2>0$) is followed by a "sweep" of fast fluid moving back downward towards the wall ($u_1>0$, $u_2<0$).

We expressed above the "hope" for deeper understanding of turbulence; what goals for further understanding could one have? After all, we have a closed system of equations for incompressible flow, the Navier-Stokes (NS) equations — see (22)–(23) — which are essentially the expressions of mass and momentum conservation in the continuum limit and which are believed to be an excellent mathematical model for subsonic flows at "reasonable" temperatures and pressures. So what is the problem? It is that we cannot solve these equations in any meaningful sense for physically relevant initial conditions; in fact, a general global existence and uniqueness theory for solutions in $\mathbb{R}^3$ is still lacking (Doering & Gibbon, 1995)!

What would one seek to "explain?" There are many different scientific and technological questions pertinent to turbulence: On the one hand, one might wish to know how such complex spatial and temporally "random" motions could arise from simple local laws — this may be understood, at least qualitatively, using ideas from dynamical systems theory, which was first applied to the study of turbulence by Hopf (1948) and Ruelle & Takens (1970). On the other hand, how might one calculate technologically important quantities such as the drag or mixing rates in a turbulent flow? A large variety of experimental, computational and theoretical techniques has been brought to bear on this problem, but much remains poorly understood. It is our hope in the work described here that a well-chosen but highly simplified model, or caricature, of the system, which makes use of the presence of coherent structures to choose an appropriate basis for representing velocity fields, may enable the tools of dynamical systems theory to shed new light on the problem.

3. The Dynamical Systems Paradigm

Central to the dynamical systems viewpoint is the concept of the attractor. In brief, an attractor is an invariant, indecomposable subset of phase space which attracts points from its exterior. Characterizing a system's attractor is a big step towards understanding its dynamics. Simply knowing that an attractor exists can be quite helpful.

The seeds of our current effort were planted by Ruelle and Takens (1970). Their hope was to reduce the dynamics of the Navier Stokes (NS) equations, describing turbulent flow, to those on an attractor and study their bifurcations and dynamics in the context of finite dimensional flows and maps. The NS equations are infinite dimensional, yet in special cases, they have been shown to possess a finite dimensional attractor. It is believed that in general the NS equations possess an attractor (Temam, 1988; Doering & Gibbon, 1995). If we could choose a subspace S_n such that the attractor, $\mathcal{A}$, and possibly other orbits asymptotic to it, formed a graph over S_n, then we could perform something akin to an inertial manifold reduction[1]. Letting p and q be coordinates in S_n and $S_n^{\perp}$ respectively, the original system may be written as

$$\begin{aligned} \dot{p} &= f(p,q), \\ \dot{q} &= g(p,q). \end{aligned} \tag{1}$$

Denoting $q = h(p)$ as the graph of $\mathcal{A}$ above S_n, we can reduce the system to S_n by letting $\dot{p} = f(p, h(p))$. This reduces the dimension of the system to that of S_n. The bad news is that by "finite dimensional" in the case of the Navier Stokes equation, we mean only "less than infinite". Rough estimates on the dimension are on the order of $Re^{9/4}$ which gives $\mathcal{O}(10^6 - 10^9)$ dimensions for flows of scientific and technological interest. (Here $Re = ul/\nu$ is a Reynolds number formed from the turbulent velocity,

[1] For details of inertial manifolds, see Temam (1988). Note, however, that the question of the existence (or non-existence) of an inertial manifold for the Navier-Stokes equations, even in the two-dimensional case, has not yet been resolved.

u, the "integral" or energy containing length scale, l, and the kinematic viscosity, ν.)

Clearly a "simple" reduction to the attractor will not be sufficient to produce an ODE model of a size which might be studied analytically. Thus, we less ambitiously look for a subspace S_n which traps "most" of the dynamics, not all of it. In our investigations, we will take "most" to mean most of the kinetic energy on average. The naïve picture we have is a conditional probability measure, $\mu(q|p)$, $p \in S_n$ and $q \in S_n^{\perp}$, sitting on the fiber Q_p above each point $p \in S_n$. Such a representation assumes the existence of a finite invariant measure for the dynamics so that the conditional distribution on the fiber above a base point is fixed in time, and can be averaged out of the dynamics in a time independent fashion. The hope is that the behavior would be well described by the dynamics projected onto S_n, and augmented by the averaged effect of the neglected modes (Berkooz, 1994):

$$\dot{p} = \int_{Q_p} f(p,q)\, d\mu(q|p). \tag{2}$$

In the absence of detailed *a priori* information on invariant measures, this idea, augmented with substantial physical input, will guide our modeling of neglected effects. It seems essential to include some information on the neglected modes, in analogy with the advantages of a center manifold, compared with a linear center eigenspace truncation.

4. The Proper Orthogonal Decomposition

The existence of coherent structures in certain flows, which contain most of the energy, suggests that the dramatic reduction in dimension postulated in the previous section might be achieved by a suitable modal decomposition which retains these structures and averages or models the incoherent fluctuations. The *Proper Orthogonal Decomposition* (POD) offers a rational way for extracting such features.

The POD is a procedure for extracting a basis for a modal decomposition of functions, from an ensemble of "observations" obtained experimentally or from direct numerical simulation. It is attractive in that it is a linear procedure, based on the spectral theory of compact, self-adjoint operators, so its properties and limitations are clear. Moreover, the decomposition it affords is optimal in the precise sense described below. The POD was first introduced in the context of turbulence by Lumley (1967); in other fields it is known under various alternative names, notably the *Karhunen-Loève decomposition* (see Berkooz *et al.* (1993a) for a more general discussion, and Holmes *et al.* (1996) for a detailed description, including proofs, other applications and a historical perspective).

The basic idea is straightforward: Suppose we have an ensemble $\{\mathbf{u}^k\}$ of observations, say, of experimental measurements or numerical simulations of the turbulent velocity field, where the $\mathbf{u}^k$ belong to a Hilbert space X. Our goal is to obtain an orthogonal basis φ_j for X, so that almost every member of the ensemble can be decomposed relative to the φ_j:

$$\mathbf{u} = \sum_{j=1}^{\infty} a_j \varphi_j. \tag{3}$$

There is no *a priori* reason to distinguish between space and time in the definition and derivation of the empirical basis functions, but we ultimately wish to obtain a dynamical model for the coherent structures. Hence, here we shall seek spatial vector-valued basis functions φ_j, and subsequently determine the time-dependent scalar modal coefficients a_j via projection of the governing equations; giving the decomposition

$$\mathbf{u}(\mathbf{x}, t) = \sum_j a_j(t) \varphi_j(\mathbf{x}). \tag{4}$$

Here $\mathbf{x} \in \Omega$, the spatial domain of the experiment, and typically in fluid applications, we choose $X = [L^2(\Omega)]^3$, with an inner product defined by $(f, g) = \int_\Omega \sum_{i=1}^3 f_i g_i^* \, d\mathbf{x}$ (In fact we further restrict to the subspace of divergence free vector fields, consistent with incompressible flows). Central to the POD is the concept of the averaging operation $\langle \cdot \rangle$, associated with a probability measure μ on X, and which is assumed to commute with the spatial integral of the L^2 inner product. The operation $\langle \cdot \rangle$ may simply be thought of as an average over a number of separate experiments, or, if we assume ergodicity, as a time average over the ensemble of observations obtained at different times during a single experimental run.

The discussion that follows may be justified rigorously, through a careful analysis of the averaging operation and the operators involved in the POD — see Holmes *et al.* (1996); we present only an overview, with few proofs.

4.1. DERIVATION OF EMPIRICAL EIGENFUNCTIONS

In mathematical terms, a normalised basis element φ is optimal if the average projection of $\mathbf{u}$ onto φ is maximised; that is, we seek

$$\max_{\varphi \in X} \frac{\langle |(\mathbf{u}, \varphi)|^2 \rangle}{\|\varphi\|^2}, \tag{5}$$

where $|\cdot|$ denotes the modulus and $\|\cdot\|$ is the L^2-norm, $\|f\| = (f, f)^{1/2}$. This can be reformulated in terms of the calculus of variations, with a functional for the constrained variational problem

$$J[\varphi] = \langle |(\mathbf{u}, \varphi)|^2 \rangle - \lambda(\|\varphi\|^2 - 1); \tag{6}$$

a necessary condition for extrema is the vanishing of the functional derivative for all variations $\varphi + \varepsilon\psi \in X$:

$$\frac{d}{d\varepsilon} J[\varphi + \varepsilon\psi]|_{\varepsilon=0} = 0. \tag{7}$$

Some algebra, together with the fact that $\psi(\mathbf{x})$ is an arbitrary variation, shows that the condition (7) reduces to

$$\int_\Omega \underbrace{\langle \mathbf{u}(\mathbf{x})\mathbf{u}^*(\mathbf{x}')\rangle}_{R(\mathbf{x},\mathbf{x}')} \varphi(\mathbf{x}')\, d\mathbf{x}' = \lambda\varphi(\mathbf{x}). \tag{8}$$

We may rewrite this Fredholm integral equation of the second kind, whose kernel is the averaged autocorrelation tensor $R(\mathbf{x},\mathbf{x}') \stackrel{def}{=} \langle \mathbf{u}(\mathbf{x})\mathbf{u}^*(\mathbf{x}')\rangle$, as an operator equation, $\mathbf{R}\varphi = \lambda\varphi$. The optimal basis is thus given by the eigenfunctions of this integral equation. These are frequently called *empirical eigenfunctions*, since the basis is derived from the ensemble of observations $\{\mathbf{u}^k\}$. The operator $\mathbf{R}$ is clearly self-adjoint. Furthermore, under plausible conditions on the averaging measure μ, $\mathbf{R}$ is also compact, so that Hilbert-Schmidt theory assures us that there is a countable infinity of eigenvalues $\{\lambda_j\}$ and eigenfunctions $\{\varphi_j\}$ (which we may normalise so that $\|\varphi_j\| = 1$) given by solutions of (8). We may order the eigenvalues so that $\lambda_j \geq \lambda_{j+1}$, and note that the positive semi-definiteness of R implies that $\lambda_j \geq 0$. We thus have a diagonal decomposition of the autocorrelation function:

$$R(\mathbf{x},\mathbf{x}') = \sum_{j=1}^{\infty} \lambda_j \varphi_j(\mathbf{x})\varphi_j^*(\mathbf{x}'). \tag{9}$$

It is these empirical eigenfunctions that we use in the modal decomposition (4) above. We note that the diagonal representation (9) of the two-point correlation tensor ensures that the modal amplitudes are uncorrelated:

$$\langle a_i a_j^* \rangle = \delta_{ij}\lambda_j. \tag{10}$$

In applications, we wish only to retain eigenfunctions φ_j with strictly positive eigenvalues λ_j, that is, those spatial structures having finite energy on average. It is natural to explore the nature of the span of these $\{\varphi_j\}$, that is, $S = \{\sum a_j\varphi_j | \lambda_j > 0,\ \sum |a_j|^2 < \infty\}$: which functions can be reproduced by convergent linear combinations of these empirical eigenfunctions? It turns out that almost every member of the original ensemble $\{\mathbf{u}^k\}$ (with respect to the probability measure underlying the averaging operation) belongs to S. Conversely, the elements of S (including the empirical eigenfunctions) may be constructed in terms of the $\{\mathbf{u}^k\}$:

$$\theta \in S \Rightarrow \theta(\mathbf{x}) = \sum_k b_k \mathbf{u}^k(\mathbf{x}). \tag{11}$$

The span of the eigenfunctions is thus exactly the span of all the realizations of $\mathbf{u}(\mathbf{x})$, with the exception of a set of measure zero. Frequently the realizations of $\mathbf{u}(\mathbf{x}, t)$ do not span X (much, but not *everything* happens in turbulence); in this case, the POD eigenfunctions $\{\varphi_j | \lambda_j > 0\}$ do not form a complete basis of $[L^2(\Omega)]^3$, but rather span only the smallest linear subspace that is sufficient to describe the observed phenomena — "you can only describe what you've seen already." Inclusion of all generalized eigenfunctions of $\mathbf{R}$ with zero eigenvalues gives a complete basis, but in doing so one loses the major advantage of the POD: the possibility of a drastic reduction in the dimensionality of the system.

Note that from (11), properties common to all the $\mathbf{u}$ in the ensemble are inherited by the empirical eigenfunctions. In our example, this means in particular that the φ_j satisfy the same linear boundary conditions as the velocity field, and are divergence free. As we shall see below, symmetries also pass on to the φ_j.

4.2. OPTIMALITY

Suppose we have an ensemble member $\mathbf{u}(\mathbf{x}, t)$, decomposed with respect to an (arbitrary) orthonormal basis $\{\psi_j\}$,

$$\mathbf{u}(\mathbf{x}, t) = \sum_j b_j(t) \psi_j(\mathbf{x}). \tag{12}$$

Using the orthonormality of the ψ_j, from the above expression, twice the average kinetic energy per unit mass over the experiment is given by

$$\int_\Omega \langle \mathbf{u}\mathbf{u}^* \rangle \, d\mathbf{x} = \sum_j \langle b_j(t) b_j^*(t) \rangle. \tag{13}$$

Hence, $\langle b_j b_j^* \rangle$ represents twice the average kinetic energy in the j-th mode. It follows that for the particular case of the POD decomposition (4), the energy in the j-th mode is λ_j, as indicated by (10).

We may now state an optimality result for the POD: For any N, the energy in the first N modes in a proper orthogonal decomposition is at least as great as that in any other N-dimensional projection:

$$\|\mathbf{u}_N\|^2 = \sum_{j=1}^N \langle a_j a_j^* \rangle = \sum_{j=1}^N \lambda_j \geq \sum_{j=1}^N \langle b_j b_j^* \rangle. \tag{14}$$

This follows from a result on general linear self-adjoint operators: the sum of the first n eigenvalues of $\mathbf{R}$ is greater than or equal to the sum of the diagonal terms in any n-dimensional projection of $\mathbf{R}$. Equation (14) implies that, among all *linear* decompositions, the POD is the most efficient for modeling or reconstructing a signal $\mathbf{u}(\mathbf{x}, t)$, in the sense of *capturing, on average, the most kinetic energy possible for a projection on a given number of modes*. This observation furnishes the motivation for the use of the POD in low-dimensional modeling of coherent structures. The rate of decay of the λ_j gives an indication of how fast the low-dimensional representation converges on average, and hence how well it might do for these structures.

One can define optimal bases with respect to norms other than the L^2 norm, of course. For instance, an H^1 Sobolev norm would give a basis better adapted to capturing vorticity $(\nabla \times \mathbf{u})$, an H^2 norm one more geared to dissipation $(\nabla^2 \mathbf{u})$, etc.

4.3. SYMMETRIES IN THE P.O.D.

As indicated above, the empirical eigenfunctions inherit properties shared by all observations $\mathbf{u}^k$, including symmetries. For our purposes, the most important symmetry is *homogeneity*. This is the condition that the averaged two point correlation be translation invariant, depending only on the difference between coordinates: $R(x, y) = R(x - y)$. It is generally a property of the system as a whole, which occurs in spatially unbounded as well as periodic systems. Furthermore, a system may be homogeneous in some directions and not in others. Also, while the ensemble of realisations $\{\mathbf{u}^k\}$ may be translation invariant on average, individual realisations typically are not.

This property is useful because in the homogeneous directions, the empirical eigenfunctions are simply Fourier modes: $\varphi_k(x) \propto e^{ikx}$. This is readily verified by substituting this *Ansatz* for the eigenfunctions into the definition (8) in the homogeneous directions, for an unbounded or periodic domain Ω:

$$\begin{aligned} \int_\Omega R(x-y) e^{iky}\, dy &= \int_\Omega R(\eta) e^{ik(x-\eta)}\, d\eta \\ &= \left(\int_\Omega R(\eta) e^{-ik\eta}\, d\eta \right) e^{ikx} \\ &\stackrel{def}{=} \tilde{\lambda}_k e^{ikx}. \end{aligned} \tag{15}$$

Hence e^{ikx} is indeed an eigenfunction, with eigenvalue $\tilde{\lambda}_k$; the eigenvalues being given by the Fourier transform of the averaged autocorrelation. (Note that this indexing of the eigenvalues, by their Fourier mode number, does not necessarily correspond to the ordering in decreasing magnitude which we introduced earlier, as analysis of the spectrum may show that a mode higher than the first contains the most energy; this occurs for instance in the Kuramoto-Sivashinsky equation, to be discussed in section 10, and in the spanwise direction of the boundary layer — see figure 4. For consistency with our previous notation, we may re-order by introducing a permutation, $i \to k(i)$, so that $\lambda_i \equiv \tilde{\lambda}_{k(i)}$, and $\lambda_i \geq \lambda_{i+1}$.)

The *continuous* symmetry of homogeneity, or translational invariance, is only one of many symmetries that fluids may exhibit, depending on the experimental setup. A given system could be invariant under *discrete* symmetries, such as translations through multiples of some finite length, or discrete rotations, for example. Such discrete symmetries would occur in a boundary layer treated with riblets — strakes parallel to the mean flow direction. We shall later investigate in greater depth the effect of such symmetries on the dynamical behavior of the system.

4.4. APPROXIMATING ATTRACTORS

In implementing the POD for low-dimensional modeling, we project the governing infinite-dimensional evolution equation, such as the NS equations, onto a finite-dimensional empirical subspace, of possibly quite low dimension. The question naturally arises: How well does this truncation and projection approximate the attractor of the original dynamical system?

For insight into this question, we appeal to an elementary inequality from probability theory, which gives an upper bound on the frequency of departures from the mean:

CHEBYSHEV'S INEQUALITY: *Let* $\mathbf{x}$ *be a vector-valued random variable, with mean* $\langle \mathbf{x} \rangle$ *and variance* $\sigma^2 = \langle |\mathbf{x} - \langle \mathbf{x} \rangle|^2 \rangle$. *Then for any* $\varepsilon > 0$,

$$P\{|\mathbf{x} - \langle \mathbf{x} \rangle| \geq \varepsilon\} \leq \frac{\sigma^2}{\varepsilon^2}. \tag{16}$$

We use this inequality to estimate the probability of the system evolution remaining close to a finite-dimensional subspace spanned by the first n empirical eigenfunctions, $S_n = span\{\varphi_1, \ldots, \varphi_n\}$. The projection onto the first n modes belongs to S_n; to estimate the error incurred by neglecting the remaining modes, we look at the infinite "tail." To do this, we define an infinite-dimensional "slab" of thickness 2ε around S_n,

$$S_n(\varepsilon) = \{u \in L^2(\Omega) | \sum_{m=n+1}^{\infty} |(u, \varphi_m)|^2 < \varepsilon\},$$

and let $W_n(\varepsilon) = L^2(\Omega) \backslash S_n(\varepsilon)$ denote the rest of phase space. To apply Chebyshev's inequality to estimate the fraction of time spent in $S_n(\varepsilon)$ by solutions $\mathbf{u}(\mathbf{x},t) = \sum_m(\mathbf{u},\varphi_m)\varphi_m = \sum_m a_m\varphi_m$, we let $\mathbf{y}_n$ denote the vector-valued random process

$$\mathbf{y}_n(t) = \{(\mathbf{u},\varphi_m)\}_{m=n+1}^{\infty}.$$

Then we have $\langle \mathbf{y}_n \rangle = 0$ and $\sigma^2(\mathbf{y}_n) = \sum_{m=n+1}^{\infty}\langle a_m a_m^* \rangle = \sum_{m=n+1}^{\infty}\lambda_m$ (from (10)), and therefore, by Chebyshev's inequality,

$$P\{\mathbf{u} \notin S_n(\varepsilon)\} = P\{\mathbf{u} \in W_n(\varepsilon)\} = P\{|\mathbf{y}_n| \geq \varepsilon\} \leq \frac{\sum_{m=n+1}^{\infty}\lambda_m}{\varepsilon^2}. \quad (17)$$

Now this inequality tells us how good a given finite-dimensional truncation is, but does not yet indicate whether the solution converges onto an attractor; clearly, the expression is meaningless as $\varepsilon \to 0$ for fixed n. To get a useful convergence result, we need an estimate of the rate of decay of the residual eigenvalues in the tail, λ_n. It turns out that there is much evidence, both physical and mathematical, that the residual decays at least exponentially fast asymptotically (in the far dissipative range). The mathematical basis for this is *Gevrey regularity* of solutions of the evolution equations — see Foias & Temam (1989) for results on the Navier-Stokes equations — and the uniform exponential boundedness of the appropriate averages used in the POD — *cf.* Berkooz (1991). These results imply that, in the limit, $\lambda_n = o(\exp(-cn))$, $c > 0$; of course, these are only asymptotic results, not directly applicable to the finite and low-dimensional modes of interest to us, where we truncate far below the far dissipative range, but they do provide a guide.

Equipped with the exponential decay of the empirical eigenvalues, we may then choose a sequence of $\varepsilon_n \to 0$ so that

$$\frac{\sum_{m=n+1}^{\infty}\lambda_m}{\varepsilon_n^2} \to 0 \quad \text{as } \varepsilon_n \to 0;$$

for instance, we may take $\varepsilon_n = o(\exp(-dn))$, for some $d < c/2$. This choice gives us a sequence of slabs $S_n(\varepsilon)$ with thickness going exponentially to zero, while the probability that the solution is in the slab, goes exponentially to one. In this picture, we have an attractor which is very thin, even though very high or even infinite-dimensional. The thin slabs capture more and more of the energy as we include more modes, and it is reasonable to postulate that the dynamics is controlled by a finite number of modes; this forms the basis for our work. Note that the above attractor may not be compact; however, it is possible to show that with exponential decay of the empirical eigenvalues, practically all the support of the invariant measure of the dynamics *is* contained in a compact set. For a further introductory discussion of these regularity results, see Berkooz *et al.* (1993a); a fuller discussion of the POD, describing several other properties and applications, is given in Holmes *et al.* (1996).

5. Representation of Boundary Layer Flows

The previous section provided a theoretical foundation for projection onto an orthonormal set of empirical eigenfunctions. How is this carried out in practice?

A set of data is required in order to obtain the averaged autocorrelation function $R(\mathbf{x}, \mathbf{x}')$, whose eigenfunctions form the POD basis. The data could be experimental or numerical; the larger the data set, the better the statistics. The original study of Aubry *et al.* (1988) used eigenfunctions derived from experiments conducted in a circular pipe at Reynolds number $Re \sim 6\,750$, by Herzog (1986) at Penn State. Later studies used data sets obtained numerically by Moin & Moser (1989; also Moin, 1984), for direct numerical simulations of channel flow at $Re \sim 3$–$4\,000$, and large eddy simulations at $Re \sim 13\,800$.

5.1. SYMMETRIES OF THE EMPIRICAL EIGENFUNCTIONS

All of these data sets exhibit certain symmetries, which consequently carry over to the empirical eigenfunctions. In particular, the fully developed boundary layer, in either channel or pipe, is translation invariant[2] (on average) in the streamwise and spanwise directions. In the light of the discussion in the previous section, this implies that the POD eigenfunctions in those directions are Fourier modes, and may be labelled by their Fourier wavenumbers. Indeed, translation invariance implies that

$$R(x_1, x_1', x_2, x_2', x_3, x_3') = R(x_1 - x_1', x_2, x_2', x_3 - x_3'),$$

where x_1 denotes the streamwise, x_2 the wall normal and x_3 the spanwise direction (see figure 3). We may thus work with the power spectrum, and take the (discrete) Fourier transform of R in the 1 and 3 directions (over domains of lengths L_1 and L_3 respectively) to give $S(k_1, k_3; x_2, x_2')$. It is S whose eigenfunctions provide the empirical basis, $\phi_n(k_1, k_3; x_2) \equiv \phi^{(n)}_{k_1 k_3}(x_2)$, the corresponding eigenvalues $\lambda^{(n)}_{k_1 k_3}$ describing the average energy content in each eigenfunction family (n), at each Fourier wavenumber pair (k_1, k_3). The domain size in the normal direction is X_2, and the eigenfunctions satisfy the orthonormality condition

$$(\phi^{(m)}_{k_1 k_3}, \phi^{(n)}_{k_1 k_3}) = \int_0^{X_2} \phi^{(m)}_{i_{k_1 k_3}}(x_2) \phi^{(n)*}_{i_{k_1 k_3}}(x_2)\, dx_2 = \delta_{mn},$$

[2] In the case of a pipe, "spanwise translation" corresponds to rotation about the axis.

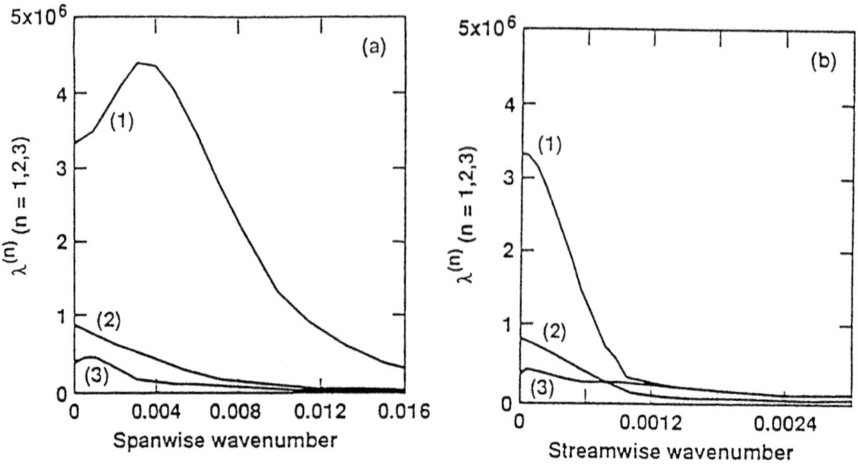

Figure 4. Empirical eigenvalue spectra for the near wall region: $\lambda^{(n)}$ as a function of (a) spanwise wavenumber k_3/L_3, (b) streamwise wavenumber k_1/L_1; from (Herzog, 1986).

where ϕ_i is the i-th component of $\boldsymbol{\phi}$, and summation over i is assumed.

Figure 4 shows the dependence of the empirical eigenvalues on spanwise Fourier wavenumber, derived from the data of Herzog (1986). The peak corresponds to the average streak spacing seen in figure 2. In the streamwise direction, the analogous spectrum is peaked at $k_1 = 0$, indicating considerably longer scales in this direction. We will refer to this empirical eigenvalue spectrum in choosing specific truncations, below.

In addition to translation invariance, the eigenfunctions $\boldsymbol{\phi}(x_2)$ inherit other symmetries from those in the experimental or numerical two-point correlation tensor (see section 4.3). In particular, when taken together with the Fourier modes, the eigenfunctions are divergence-free; furthermore, there is reflection symmetry about a midplane in the spanwise direction, and the entries of the tensor are real. In terms of the eigenfunctions, these last two symmetries imply

$$\begin{aligned} \boldsymbol{\phi}^{(n)}_{-k_1 k_3} &= (\phi^{(n)*}_{1_{k_1 k_3}}, \phi^{(n)*}_{2_{k_1 k_3}}, -\phi^{(n)*}_{3_{k_1 k_3}}), \\ \boldsymbol{\phi}^{(n)}_{k_1 -k_3} &= (\phi^{(n)}_{1_{k_1 k_3}}, \phi^{(n)}_{2_{k_1 k_3}}, -\phi^{(n)}_{3_{k_1 k_3}}), \\ \boldsymbol{\phi}^{(n)}_{-k_1 -k_3} &= (\phi^{(n)*}_{1_{k_1 k_3}}, \phi^{(n)*}_{2_{k_1 k_3}}, \phi^{(n)*}_{3_{k_1 k_3}}). \end{aligned} \tag{18}$$

Consistent with the observations and empirical eigenfunctions derived from them, the NS equations (with periodic boundary conditions in the x_1 and x_3 directions) themselves display the same symmetries, namely equivariance under translations and reflections in the spanwise direction and translation in the streamwise direction:

$$x_3 \mapsto x_3 + \xi_3; \;\; x_3 \mapsto -x_3; \;\; x_1 \mapsto x_1 + \xi_1. \tag{19}$$

In terms of the eigenfunction expansion (4), the eigenfunctions $\varphi_n(\mathbf{x})$ thus consist of Fourier modes in the x_1 and x_3 directions, and the above functions $\phi_{k_1k_3}^{(n)}$ in the x_2 direction. The index n represents a *family* of empirical eigenfunctions, for different values of k_1 and k_3 (the energy of a family of eigenfunctions is derived by summation over the wavenumbers k_1 and k_3). The full expansion, in the form (4), for the i-th component of the velocity field (where $i = 1, 2, 3$) is thus

$$u_i(x_1, x_2, x_3, t) = \frac{1}{\sqrt{L_1 L_3}} \sum_{n=1}^{\infty} \sum_{\substack{k_1=-\infty \\ k_3=-\infty}}^{\infty} a_{k_1k_3}^{(n)}(t) e^{2\pi i(\frac{k_1}{L_1}x_1 + \frac{k_3}{L_3}x_3)} \phi_{i_{k_1k_3}}^{(n)}(x_2). \tag{20}$$

The above symmetry (18) implies for the modal coefficients

$$a_{k_1k_3}^{(n)} = a_{-k_1-k_3}^{(n)*}, \tag{21}$$

guaranteeing reality of the velocity field.

After substituting (20) for the velocity field into the Navier-Stokes equations, truncating at some finite number of modes $n = N$, $k_1 = \pm K_1$, $k_3 = \pm K_3$, and taking the inner product with each included mode in turn, we obtain the desired dynamical system, a finite system of ODEs for the amplitudes $a_{k_1k_3}^{(n)}(t)$. It is this system that we derive in the next section, and whose properties we shall subsequently study. We will consider specific choices for length scales L_j and truncation wavenumbers K_j, N in section 7.

6. Projection of the Navier-Stokes Equations and Modeling

We first rewrite the NS equations in a way that isolates the dynamics of coherent structures and their interaction with the mean flow; both the neglected (high) wavenumber modes and the mean flow are modeled. Next, we project the NS equations onto the empirically obtained family of subspaces, using a Galerkin projection. Lastly, we need to truncate the system in some way — indeed, rather drastically — in order to obtain a finite and low-dimensional system that is amenable to analysis and understanding. Several assumptions are necessary in the projection process, and they will be touched on and justified here briefly; the reader should refer to Holmes *et al.* (1996) for more details.

We begin with the incompressible Navier-Stokes equations without body forces, for the fluid velocity field $\mathbf{v}(\mathbf{x}, t)$ and pressure field $\pi(\mathbf{x}, t)$:

$$\mathbf{v}_t + \mathbf{v} \cdot \nabla \mathbf{v} = -\frac{1}{\rho} \nabla \pi + \nu \Delta \mathbf{v}, \tag{22}$$

$$\nabla \cdot \mathbf{v} \;=\; 0. \tag{23}$$

We choose periodic boundary conditions in the streamwise and spanwise directions, that is on $x_1 = 0,\ L_1$ and $x_3 = 0,\ L_3$, no slip boundary conditions on the wall, that is $\mathbf{v} = 0$ on $x_2 = 0$, and an as yet unknown boundary condition $\mathbf{v} = \mathbf{f}(x_1, x_3, t)$ on the upper boundary of the wall layer $x_2 = X_2$.

6.1. REYNOLDS DECOMPOSITION

We perform the Reynolds decomposition of the fluid into a mean flow and a fluctuating field, $\mathbf{v} = \mathbf{U} + \mathbf{u}$, $\pi = P + p$, where $\mathbf{U} \equiv \langle \mathbf{v} \rangle_s$ and $P \equiv \langle \pi \rangle_s$ are spatially averaged quantities in the x_1, x_3 directions; for instance:

$$\mathbf{U}(x_2, t) = \langle \mathbf{v} \rangle_s = \frac{1}{L_1 L_3} \int_0^{L_1} \int_0^{L_3} \mathbf{v}(x_1, x_2, x_3, t)\, dx_1 dx_3.$$

If L_1 and L_3 are large enough, the mean flow $\mathbf{U} = U(x_2, t)\mathbf{e}_1$ is in the streamwise x_1 direction, and has a slow enough time dependence that we can neglect $\mathbf{U}_t$ relative to $\mathbf{u}_t$.

Averaging (22), we obtain the equation for the mean flow,

$$\mathbf{U}_t + \mathbf{U} \cdot \nabla \mathbf{U} + \langle \mathbf{u} \cdot \nabla \mathbf{u} \rangle_s + \frac{1}{\rho} \nabla P - \nu \Delta \mathbf{U} = 0,$$

where $\langle \mathbf{u} \cdot \nabla \mathbf{u} \rangle_s$ is the Reynolds stress term. Using $\mathbf{U} \cdot \nabla \mathbf{U} = 0$, $\mathbf{U}_t \approx 0$, and assuming that streamwise and spanwise variations of mean quantities are negligible, we obtain for the mean flow, in component form,

$$\langle u_{i,j} u_j \rangle_s = -\frac{1}{\rho} P_{,i} + \nu U_{,22} \delta_{i1}. \tag{24}$$

Subtracting the mean equation (24) from (22), we obtain an evolution equation for the turbulent fluctuations,

$$N(\mathbf{u}) = \mathbf{u}_t + \mathbf{U} \cdot \nabla \mathbf{u} + \mathbf{u} \cdot \nabla \mathbf{U} + \mathbf{u} \cdot \nabla \mathbf{u} - \langle \mathbf{u} \cdot \nabla \mathbf{u} \rangle_s + \frac{1}{\rho} \nabla p - \nu \Delta \mathbf{u} = 0. \tag{25}$$

6.2. THE MEAN FLOW

In order to close the equations for $\mathbf{u}$, we need to model the mean flow in some way. In Aubry *et al.* (1988), the mean flow was expressed as a balance between the effects of pressure and those of the coherent structures. Order of magnitude estimates, which may be justified by asymptotic calculations through manipulation of the first and second components of (24), give (see Holmes *et al.* (1996) or Berkooz *et al.* (1993b))

$$U(x_2,t) = \frac{1}{\nu}\int_0^{x_2} \langle u_1 u_2\rangle_s(x_2',t)\,dx_2' \;+\; \frac{u_\tau^2}{\nu}\left(x_2 - \frac{x_2^2}{2H}\right). \tag{26}$$

Here the first term on the right-hand side is due to the feedback from the turbulent bursts into the "locally averaged" mean flow, while the second term is a "constant" driving term corresponding to the mean pressure gradient. u_τ is the friction velocity giving a measure of the shear stress at the wall, defined by $u_\tau^2 = \nu \frac{\partial U}{\partial x_2}|_{x_2=0}$, and we assume $x_2 \ll H$, where H is half the width of the channel. Substitution of (26) into the full equations (25) gives a closed system amenable to further analysis; note that the first term in the expression for the mean flow will give rise to a cubic term in the full system, which implies global stability for our dynamical system.

6.3. LOSSES TO NEGLECTED MODES

The next step in our analysis is to take account of the fact that we will be performing a rather violent truncation, only retaining modes that are linearly unstable or weakly stable, without keeping "stable" modes that are nevertheless dynamically active. With this drastic truncation the energy cascade mechanism that forms the basis for the turbulent energy transfer from long to short wavelength modes is lost, and must be modeled; otherwise the energy balance, in which the small-scale turbulent motions act as energy sinks for the coherent structures, is lost. To obtain such a model, we decompose the turbulent velocity field into resolved components $\mathbf{u}_<$, representing the coherent structures, and unresolved smaller scale components $\mathbf{u}_>$: $\mathbf{u} = \mathbf{u}_< + \mathbf{u}_>$; and we project equation (25) onto the subspace of resolved components. Once again, in order to have a closed system, we need to model the small-scale $\mathbf{u}_>$ in terms of the larger scales $\mathbf{u}_<$. For this, we use a Heisenberg spectral transfer model, or eddy viscosity mechanism; for details, see Holmes *et al.* (1996, ch.9). In the modeling, two "free" parameters α_1 and α_2 are introduced, which should both be of order one, but which can be adjusted to obtain the appropriate energy flow to the unresolved modes. These parameters (which, for simplicity, are later equated) lead to a bifurcation parameter in our dynamical system model.

Finally, before writing down the restricted partial differential equation for the resolved modes, we normalize the equations using wall units. That is, using the friction velocity u_τ and viscosity ν, the units of length, time and velocity are ν/u_τ, ν/u_τ^2 and u_τ, respectively; the unit of pressure is ρu_τ^2, and we denote normalized pressure by p^τ. In such wall variables, all wall bounded flows with the same pressure gradient look the same; in particular, the Reynolds number in wall units is unity, so that viscosity does not appear in the normalized equations.

6.4. GALERKIN PROJECTION

The net result of modeling the mean flow and the neglected modes is a partial differential equation for the resolved modes — the coherent structures:

$$\begin{aligned}\frac{\partial u_i}{\partial t} &= -u_{i,1}\left[\int_0^{x_2} \langle u_1 u_2\rangle \, dx_2' + (x_2 - \frac{x_2^2}{2H})\right] \\ &\quad -u_2\,\delta_{i,1}\left[\langle u_1 u_2\rangle + (1-\frac{x_2}{H})\right] \\ &\quad +(1+\alpha_1\nu_T)u_{i,jj} - [(u_{i,j}u_j) - \langle u_{i,j}u_j\rangle] \\ &\quad +\frac{1}{3}\alpha_2 l_>^2[(u_{k,l}+u_{l,k})(u_{k,l}+u_{l,k}) \\ &\quad -\langle (u_{k,l}+u_{l,k})(u_{k,l}+u_{l,k})\rangle]_{,i} - p_{,i}^{T}. \end{aligned} \tag{27}$$

Here ν_T and $l_>$ are a viscosity and length scale characteristic of the unresolved modes, which are model parameters implicitly dependent on the domain size and wave number cutoff; we omit the subscript $<$ on $\mathbf{u}_<$.

We now perform the Galerkin projection onto the empirical subspace S_n, substituting the modal decomposition (20) into (27). In this there are two steps: Fourier transformation in the x_1, x_3 directions, followed by taking the inner product with the empirical eigenfunction $\phi_{k_1 k_3}^{(n)}(x_2)$. The details for the Galerkin projection, and the considerable algebra involved, are omitted here — see Holmes *et al.* (1996) for details and for expressions and numerical values for the coefficients — but only the results are sketched. The process of Galerkin projection will be described in more depth for the far simpler case of the Kuramoto-Sivashinsky equation in section 10.

After some work, we obtain a system of the form

$$\begin{aligned}\dot{\mathbf{a}} &= [\mathbf{A}_1 + (1+\alpha_1\nu_T)\mathbf{A}_2]\mathbf{a} \\ &\quad + [\mathbf{Q}_1(\mathbf{a},\mathbf{a}) + \frac{2}{3}\alpha_2 l_>^2 \mathbf{Q}_2(\mathbf{a},\mathbf{a})] + \mathbf{C}(\mathbf{a},\mathbf{a},\mathbf{a}) + \xi(t), \end{aligned} \tag{28}$$

where $\mathbf{a}$ denotes the vector of modal coefficients $a_{k_1,k_3}^{(n)}(t)$, and the terms $\mathbf{A}$, $\mathbf{Q}$ and $\mathbf{C}$ are, respectively, linear, quadratic and cubic terms in $\mathbf{a}$.

6.5. THE PRESSURE TERM

Of particular interest here is the term $\xi(t)$, which represents a boundary pressure term. In typical projections of the Navier-Stokes equations, the pressure gradient is eliminated by integration by parts, but in our case an unusual boundary condition is being used in the wall normal direction — a spatial cutoff at $X_2 = 40$ wall units, quite near the wall. As noted in section 2, we want to model only the wall region, and so some such cut-off or weight function must be used.

We briefly indicate how the term $\xi(t)$ arises: By the divergence theorem, we have the identity for the inner product between the gradient of a scalar, ∇p, and a divergence-free vector $\mathbf{u}$,

$$(\nabla p, \mathbf{u}) = \int_\Omega \nabla p \cdot \mathbf{u}\, d\mathbf{x} = \int_{\partial\Omega} \mathbf{n} \cdot (p\mathbf{u})\, dS.$$

In our particular case, after Fourier transformation and projection onto the empirical eigenfunctions, which are also divergence-free, we have

$$\begin{aligned}(\nabla p, \boldsymbol{\phi}) &= \int_0^{L_1}\int_0^{X_2}\int_0^{L_3} \nabla p \cdot \boldsymbol{\phi}\, dx_3\, dx_2\, dx_1 \\ &= \int_0^{X_2}\int_0^{L_3} p\phi_1\, dx_3\, dx_2 \Big|_0^{L_1} + \int_0^{L_1}\int_0^{L_3} p\phi_2\, dx_3\, dx_1 \Big|_0^{X_2} \\ &+ \int_0^{L_1}\int_0^{X_2} p\phi_3\, dx_2\, dx_1 \Big|_0^{L_3}. \end{aligned} \tag{29}$$

The first and third term on the right hand side cancel identically because of periodic boundary conditions in the x_1, x_3 directions, while the middle term at $x_2 = 0$ also vanishes, since the empirical eigenfunctions satisfy the no-slip condition at the wall. However, the remaining term survives, as the empirical eigenfunctions do not vanish on the upper edge of the wall region; the pressure term thus transforms as

$$p^\tau \rightarrow \int_0^{L_1}\int_0^{L_3} p\phi_2\, dx_3\, dx_1 \Big|_{x_2 = X_2} = \hat{p}^\tau_{k_1 k_3}(X_2)\phi^{(n)*}_{2_{k_1 k_3}}(X_2). \tag{30}$$

The conditions at this boundary are not specified within the model, but must be estimated or supplied from independent computations. However, the presence of this pressure term is a potentially useful feature, since the pressure field at $x_2 = X_2$ represents communication between the modeled wall layer and the outer region of the flow; as such, one can compare results between when it is present and absent, and hence distinguish between inner and outer influences on the dynamics in the wall layer. (Note that the quadratic pseudo-pressure term $\mathbf{Q}_2$ also arises due to the "free surface", and would vanish otherwise; however, it is small compared to $\mathbf{Q}_1$ and its effects may often be neglected.)

A simple approach to modeling the pressure field is simply to use small additive random excitations characteristic of disturbances in the outer layer;

thus we have the random term $\xi(t)$. Order of magnitude estimates indicate that typically $|\xi_i/a_i| \sim \mathcal{O}(10^{-2})$, so that to a first approximation one can ignore the additive random term. Later in section 11, the effect of noise will be included, and we shall be able to distinguish between inner and outer layer effects and see that the term $\xi(t)$ may play an important rôle.

7. Structure and Some Solutions of the Models

In the previous section, we outlined the Galerkin projection, which yields a system of ordinary differential equations from the NS equations. This process in fact produces a nested hierarchy of models, containing more or fewer modes, as one seeks to approximate a greater or lesser fraction of the average energy by choosing L_1, L_3, X_2 and K_1, K_3, N in the truncated representation (20) appropriately. One must balance the inclusion of more modes, which implies a better representation of the velocity field (in terms for instance of energy captured), with increasing dimension, which renders the ODEs more unwieldy, analytically complicated and opaque. Ideally, we would like to study systems that are as small as possible, but will still yield relevant results.

7.1. CHOICE OF TRUNCATION

With this in mind, the original system studied (Aubry *et al.*, 1988) was a minimal truncation which incorporated only a single family of eigenfunctions ($N = 1$) and a single streamwise Fourier mode ($K_1 = 0$); that is, the velocity fields have no streamwise dependence. In the spanwise direction, up to $K_3 = 8$ modes were initially included, although almost all studies were done with $K_3 = 5$. The reasons for this choice and the choice of length scales L_1, L_3 and X_2, were the following:

We wish to reproduce the cross-stream interactions that contribute to the observed turbulent bursts. It appears that a spanwise domain length of $L_3 = 333$ and five nonzero modes ($K_3 = 5$) can achieve this, being large enough to permit the interaction between two eddy pairs that lead to bursts. One must certainly include wavenumbers characteristic of the spanwise energy spectrum exhibited by the empirical eigenvalues, with one near the peak and enough "interstitial" wavenumbers to afford realistic interactions; see figure 4.

In the streamwise direction, the eigenvalue spectrum as a function of streamwise wavenumber peaks at the origin, indicating a long recurrence period for coherent structures in the streamwise direction. It thus seems plausible to consider at first coherent structures with no streamwise variation, and it hence does not matter how long the domain is in that direction ($L_1 = 333$ was chosen by Aubry *et al.* (1988)).

In the direction normal to the wall, a distance of $X_2 = 40$ in wall units was chosen; in this direction, there are the conflicting requirements of resolving as much of the coherent structures, which extend well into the flow, as possible, while also not requiring excessive complexity — further from the wall, the range of spatial scales and consequently the number of excited degrees of freedom increases; see Moin (1984), for example. The above choice is a compromise, which permits us to retain just one family of empirical eigenfunctions for now.

Having chosen K_1, K_3 and N, we may write down the modal equations for the velocity field which, being constant in the streamwise direction, we may term "two-and-a-half dimensional." The decomposition (20) reduces to

$$\mathbf{u}(\mathbf{x},t) = \sum_{k_3=-K_3}^{K_3} a^{(1)}_{0,k_3}(t) e^{2\pi i \frac{k_3}{L_3} x_3} \phi^{(1)}_{0,k_3}(x_2), \tag{31}$$

and the ODEs simplify considerably; in component form they are

$$\begin{aligned} \dot{a}_k &= [A^1_k + (1+\alpha\nu_T)A^2_k]a_k \\ &\quad + [\sum_{k'} Q^1_{k\ k-k'} a_{k'} a_{k-k'} + \frac{2}{3}\alpha l^2_{>} \sum_{k'} Q^2_{k\ k-k'} a_{k'} a_{k-k'}] \\ &\quad + (\sum_{k'} C_{k\ k'} |a_{k'}|^2) a_k + \xi_k(t), \end{aligned} \tag{32}$$

where $a_k \equiv a^{(1)}_{0,k_3}$, and we have set the loss parameters $\alpha_1 = \alpha_2 = \alpha$. That is, in this special case, the linear terms become diagonal, simplifying the analysis considerably, and the cubic terms also take a simple form. In fact the overall structure is similar to that of the Fourier mode projections of the single space dimensional Kuramoto-Sivashinsky equation, to be discussed later.

The cubic terms are of particular importance here, as they guarantee stability; this is because the coefficients $C_{kk'}$ turn out to be all negative. Hence one can easily construct a Liapunov function for this system, and demonstrate the existence of an absorbing ball B for the dynamics. The attractor may then be found as the image of B asymptotically in time: $\mathcal{A} = \bigcap_{t\geq 0} \varphi_t(B)$, where φ is the dynamical evolution operator.

Specific coefficients A^j_k, $Q^j_{k\ k-k'}$, $C_{kk'}$ for the case $K_3 = 5$ are tabulated in the Appendix of Aubry *et al.* (1988); however, we note that a multiplicative factor of $\sqrt{L_1 L_3}$ was inadvertently omitted from that paper, leading to a compression of time scales by 333. This led to an erroneous interpretation of bursting timescales, which was corrected without discussion by Sanghi & Aubry (1993). We return to this issue in section 11.

7.2. SYMMETRIES

In this system, with the pressure term $\xi_k(t)$ set to zero (which we will assume for now), (32) is equivariant under the group $O(2)$ of rotations and reflections in Fourier space, corresponding to spanwise rotations by ζ_3 and reflection (see section 8 for more discussion; equivariance here simply means that elements of $O(2)$ map solutions to other solutions):

$$T_{\zeta_3} : x_3 \mapsto x_3 + \zeta_3 \text{ becomes } T_{\zeta_3} : a_k \mapsto e^{2\pi i \frac{k}{L_3}\zeta_3} a_k \tag{33}$$

$$R_f : x_3 \mapsto -x_3 \quad \text{becomes } R_f : a_k \mapsto a_k^*. \tag{34}$$

Note that our ODE is just the restriction of (28) to the subspace $k_1 = 0$; the symmetry under streamwise translations is not included explicitly here, since all non-zero streamwise modes have been excluded in (31).

By reality of the vector field, we have $a_k(t) = a_{-k}(t)$, so that we only need to solve a set of K_3 complex ODEs, or $2K_3$ real ones. For $K_3 = 5$ we thus have a 10-dimensional model system, one that is minimal but, as we shall see, quite interesting.

It is useful to look not only at symmetries, but also at invariant subspaces in this system; these provide a "backbone" for further analysis. Equation (34) implies that the purely real subspace is invariant, as is any rotation of it under T_{ζ_3} (33). Similarly, the structure of the quadratic and cubic terms of the ODEs implies that the subspace spanned by the even Fourier modes $k = 2, 4, \ldots$ is also invariant. These lower-dimensional subspaces allow us to "parse" the system and reach at least a partial understanding of its bifurcations and dynamical properties. (Note also that the $K_1 = 0$ system is invariant in higher-dimensional systems with streamwise variation having only the first family $N = 1$ of empirical eigenfunctions, so the results of the present study are useful for later work.)

7.3. BEHAVIOR OF THE MODELS

The minimal model equations (32) were studied by Aubry *et al.* (1988), and their behavior is briefly outlined here. Simulations were performed by numerical integration of the ODEs (32), and bifurcation analyses were performed on this system. The time-dependent amplitudes obtained from the integration were then substituted back into the expansion (31) to obtain the reconstructed time and space behavior of the velocity field.

A typical result of such a simulation for the modal coefficients is given in figure 5 (where x_i, y_i are the real and imaginary parts of the coefficients). The corresponding reconstructed velocity field is shown in figure 6. Note that the solution has a characteristic repeating structure, and for most of the cycle it remains near a fixed point with only the amplitudes of the 2 and

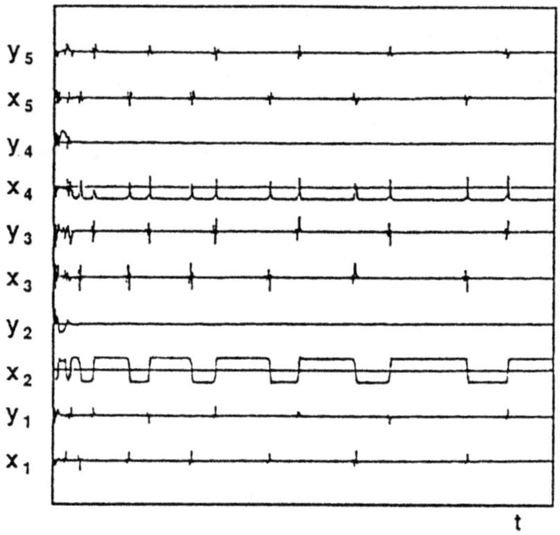

Figure 5. Time histories of the modal coefficients, for $\alpha = 1.4$; from Aubry *et al.* (1988).

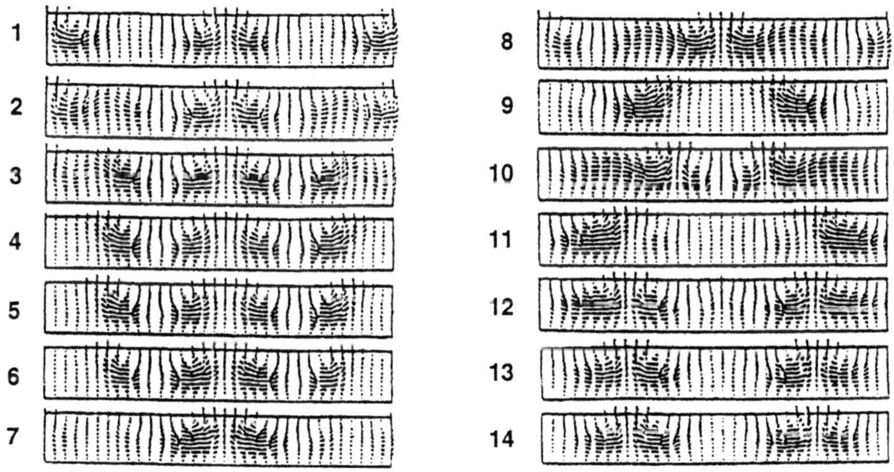

Figure 6. Cross-stream (u_2, u_3) components of the velocity field corresponding to a single heteroclinic excursion, reconstructed from the temporal dynamics of the modal coefficients at $\alpha = 1.4$, shown in figure 5. From Aubry *et al.* (1988).

4 modes nonzero; that is, the dynamics remains near the invariant subspace spanned by these modes. During this time, the amplitudes of modes 1, 3 and 5 grow exponentially in an oscillatory fashion, until eventually the 2 and 4 modes respond by changing rapidly, the phase of the 2 mode changing by π and that of the 4 mode by 2π, while the amplitudes of the 1, 3 and 5 modes simultaneously collapse to zero. The cycle then repeats. One can also see how the unstable modes switch from the real x_1, x_3, x_5 subspace to the imaginary y_1, y_3, y_5 subspace and back again. Solutions of this type persist and appear strongly attractive over a relatively wide range of parameters, α.

What is happening here? From a consideration of the behavior in phase space, we seem to have *heteroclinic cycles* — the system moves from the vicinity of one fixed point to another and back again. In fact, analyses analogous to those done later for the two mode model in section 9 reveal that the steady 2/4 mode solutions constitute a circle of saddle points with two-dimensional heteroclinic orbits connecting diametrically opposite points. Now generically, heteroclinic cycles are structurally unstable, and can be removed by a small perturbation (Guckenheimer & Holmes, 1983); yet they are observed to persist here, for general initial conditions and for ranges of parameters. How is this possible? It turns out that the symmetries of the system play a crucial rôle in stabilising the heteroclinic cycles, as we will show in the next sections.

Before proceeding to this, it is useful to consider the physical implications for the observed behavior: In the reconstructed velocity field (see figure 6) one can see clear evidence for the burst-sweep cycle that we set out to explain. That is, in spite of the extreme truncation, this 5-mode model qualitatively captures the gross features of the turbulent burst-sweep cycle in the boundary layer (including the lateral shift in the position of vortices). The evidence strongly suggests that this cycle corresponds to a heteroclinic cycle in phase space, with the mechanism driving the bursts being the excitation of an unstable mode involving spanwise wavenumbers 1, 3 and 5.

The equations (32) were simulated over a range of loss parameter values, α (high α implies high energy transfer to neglected modes). One can calculate the bifurcation diagram; see figure 7. For decreasing α, we find a globally stable equilibrium (in the 2, 4 subspace) becoming unstable to give various types of mixed mode solutions, including two "windows" of heteroclinic cycles, and travelling wave solutions. The cycles of figures 5–6 occur in window II. For smaller α, there is a complex interaction of (modulated) travelling waves and heteroclinic cycles, and solutions appear to be chaotic. Yet lower α gives simple spanwise travelling and modulated travelling waves. From the bifurcation diagram, it is also clear that the modes become unstable from the trivial equilibrium state in the order 2, 1, 3, 4, 5; this corresponds to the energy distribution in Fourier space, as revealed by the empirical eigenvalues (figure 4).

When the empirical eigenfunctions are derived from a different data set (for instance from a large eddy simulation of channel flow (Moin, 1984) rather than from Herzog's (1986) pipe flow experiments, which were used for the original computations of Aubry *et al.* (1988)) the numerical values in the bifurcation diagram change slightly, but the qualitative structure remains the same. In particular, the modes bifurcate off the trivial state in the same

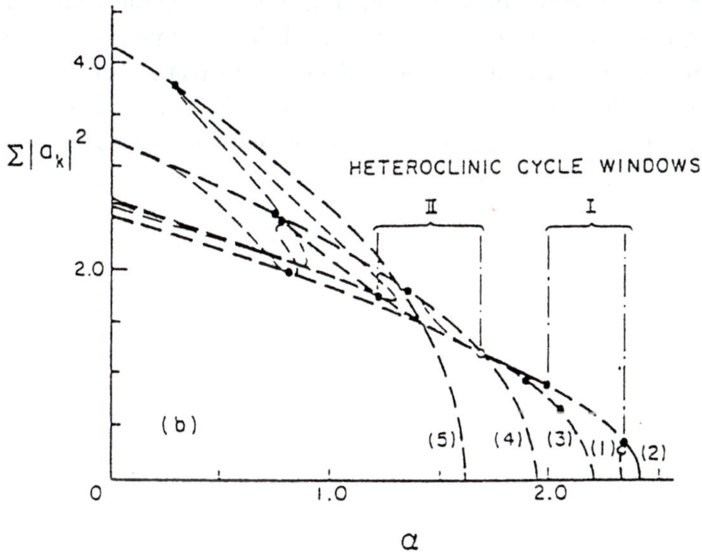

Figure 7. Bifurcation diagram showing solutions and their stability types as α is varied.

order, and α-ranges giving heteroclinic cycles still exist. This is encouraging confirmation that the picture of bursts and sweeps due to heteroclinic cycles is characteristic to boundary layer flow, not just a property of a particular data set or truncation.

In this respect, more realistic simulations, including nonzero streamwise modes and even a second family of empirical eigenfunctions, have been performed (Aubry & Sanghi, 1991; Sanghi & Aubry, 1993). Of course, the results are much more complicated, but they seem to embody the same "skeleton" of heteroclinic bursts found in the simpler minimal models, while the mechanism of the bursts, in terms of the excitation of higher modes, is captured more realistically.

Before briefly assessing these findings, we leave the boundary layer models to consider symmetry in more general terms, and to discuss some examples in which heteroclinic cycles occur in a simpler and more transparent fashion.

8. Symmetry: Translations, Reflections, and O(2)

Symmetry is a central and powerful concept in dynamics. On one hand it allows one to reduce the complexity, and often the dimension of a system. On the other hand, symmetries have profound implications for the robustness of structures in a system under perturbation. A given solution, such as homoclinic connection, may be destroyed by the smallest general perturbation. However, if only perturbations consistent with the symmetries of the system are allowed, the same solution might be structurally stable.

We have already noted how $O(2)$ appears in the boundary layer model, where it stems from translation and reflection invariance of the original physical situation (see equations (33) and (34)). To appreciate better the action of $O(2)$ on a set of Fourier modes, consider a scalar variable $w(x,t)$ and expand in a Fourier series:

$$w(x,t) = \sum z_j(t) e^{\frac{2\pi ijx}{L}}. \tag{35}$$

Translation and reflection in physical space, the x dimension, become

$$\begin{aligned} T_\alpha w(x) &= w(x+\alpha) = \sum z_j e^{\frac{2\pi ij(x+\alpha)}{L}} \\ &= \sum (e^{\frac{2\pi ij\alpha}{L}} z_j) e^{\frac{2\pi ijx}{L}}, \end{aligned} \tag{36}$$

$$\begin{aligned} R_f w(x) &= w(-x) = \sum z_j e^{\frac{2\pi ij(-x)}{L}} \\ &= \sum z_{-j} e^{\frac{2\pi ijx}{L}}. \end{aligned} \tag{37}$$

Remembering that in a physical system $w(x,t)$ is generally real, forcing $z_j^* = z_{-j}$, we conclude that translation and reflection in physical space induces the following actions on the coefficients of the Fourier modes:

$$\begin{aligned} T_\alpha : z_j &\mapsto e^{\frac{2\pi ij\alpha}{L}} z_j, \\ R_f : z_j &\mapsto z_j^* . \end{aligned} \tag{38}$$

T_α is a rotation in the complex plane and R_f is a reflection across the real axis. For z_1 this is precisely the standard action of $O(2)$ on C, the space where the coefficients live. Every member of $O(2)$ can be constructed out of a finite combination of these transformations. The T_α's together with R_f generate $O(2)$. For a general j, the only difference is that elements are rotated by an angle $j\alpha$, but this amounts to a simple rescaling. Thus, $O(2)$ represents the induced action on the phase-space C^n by the transformations (33) and (34) for any n.

In summary, if we start with a translationally invariant system and project it onto a Fourier basis, the resulting system of ODEs for the evolution of the coefficients will have an $O(2)$ symmetry. In the next section, we will take a closer look at what an $O(2)$ symmetry forces on a system. But let us begin by discussing symmetry in a slightly broader framework.

9. Equivariant ODEs and Heteroclinic Cycles

Given a group of transformations Γ on a space X (eg. $X = \mathbb{R}^n$), we say that a function $f : X \mapsto X$ is Γ-INVARIANT if

$$f(x) = f(\gamma(x)) \text{ for all } \gamma \in \Gamma. \tag{39}$$

We call a function $f : X \to X$ EQUIVARIANT if

$$\gamma(f(x)) = f(\gamma(x)) \text{ for all } \gamma \in \Gamma. \tag{40}$$

Similarly, we call a family of ODEs

$$\dot{x} = f(x) \text{ with } x \in \mathbb{R}^n \tag{41}$$

Γ-EQUIVARIANT if the vector field f is equivariant; that is,

$$\gamma(\dot{x}) = \gamma(f(x)) = f(\gamma(x)) \text{ for all } \gamma \in \Gamma. \tag{42}$$

In general, equivariant functions are built out of "pure" equivariant elements multiplied by invariant functions. As an example, consider the group D_n, the group of rotation and reflection symmetries of an n-gon in the plane. Identifying the plane, $\mathbb{R}^2$, with the complex plane, $\mathcal{C}$, for notational convenience, the D_n-equivariant functions (under the standard action on $\mathcal{C}$) are of the form

$$g(z) = p(u, v)z + q(u, v)(z^*)^{n-1},$$

with $u = zz^*$ and $v = z^n + (z^*)^n$. A quick calculation shows that $(z^*)^{n-1}$ and z are equivariant and that u and v are invariant. Thus $p(u, v)$ and $q(u, v)$ are invariant and the totality, $g(z)$ is equivariant.

9.1. $O(2)$ AND NORMAL FORMS

As already mentioned, we are mainly interested in the implications of $O(2)$ symmetry. In anticipation of the types of equations we will extract latter, we seek a set of $O(2)$-equivariant ODEs of the form

$$\begin{aligned} \dot{z}_1 &= f_1(z_1, z_2) \\ \dot{z}_2 &= f_2(z_1, z_2) \end{aligned} \quad \text{with } z_1, z_2 \in \mathcal{C}. \tag{43}$$

That is to say we seek the general form of an $O(2)$-equivariant ODE on $\mathcal{C}^2$. In this case the invariant functions turn out to depend only on the three combinations $|z_1|^2$, $|z_2|^2$ and $z_1^2 z_2^* + z_1^{*2} z_2$ and the equivariant vector fields are

$$\begin{pmatrix} z_1 \\ z_2 \end{pmatrix} \text{ and } \begin{pmatrix} z_1^* z_2 \\ z_1^2 \end{pmatrix},$$

so that, up to third order, we have the normal form

$$\begin{aligned} \dot{z}_1 &= \mu_1 z_1 + c_{12} z_1^* z_2 + (d_{11}|z_1|^2 + d_{12}|z_2|^2) z_1 \\ \dot{z}_2 &= \mu_2 z_2 + c_{11} z_1^2 \quad + (d_{21}|z_1|^2 + d_{22}|z_2|^2) z_2 \end{aligned}, \tag{44}$$

in which all the coefficients are purely real. Assuming that $c_{12}, c_{11} \neq 0$ we can make a further rescaling, reversing the direction of time if necessary, to obtain:

$$\begin{aligned} \dot{z}_1 &= z_1^* z_2 + (\mu_1 + e_{11}|z_1|^2 + e_{12}|z_2|^2)\, z_1 \\ \dot{z}_2 &= \pm z_1^2 + (\mu_2 + e_{21}|z_1|^2 + e_{22}|z_2|^2)\, z_2 \end{aligned} , \tag{45}$$

where $e_{11} = d_{11}/|c_{11}c_{12}|$, $e_{12} = d_{12}/c_{12}^2$, $e_{21} = d_{21}/|c_{11}c_{12}|$ and $e_{22} = d_{22}/c_{12}^2$. Through these rescalings, we reduce the complexity a good deal, but there still remain many cases depending on the signs of the various coefficients. Here we will analyze only the case relevant to the boundary layer model in the heteroclinic cycle range presented earlier. In this model, $e_{ij} < 0 < \mu_j$ and $\mu_1 - \mu_2 \frac{e_{12}}{e_{22}} - \sqrt{-\frac{\mu_2}{e_{22}}} < 0 < \mu_1 - \mu_2 \frac{e_{12}}{e_{22}} + \sqrt{-\frac{\mu_2}{e_{22}}}$ and we have the following proposition:

Proposition 9.1 *For $e_{ij} < 0 < \mu_j$ and $\mu_1 - \mu_2 \frac{e_{12}}{e_{22}} - \sqrt{-\frac{\mu_2}{e_{22}}} < 0 < \mu_1 - \mu_2 \frac{e_{12}}{e_{22}} + \sqrt{-\frac{\mu_2}{e_{22}}}$, the system (45) has a circle of equilibria connected by heteroclinic cycles. Furthermore, if $\mu_1 - \mu_2 \frac{e_{12}}{e_{22}} < 0$ the cycles are attracting.*

PROOF: First, observe that the z_2-plane, defined by $z_1 = 0$, is invariant. Seeking equilibria in this plane and remembering that $e_{ij} < 0 < \mu_j$, we find a circle of fixed points defined by $|z_2| = \sqrt{-\frac{\mu_2}{e_{22}}}$. Now let $z_j = x_j + iy_j$ and consider (45) as a system on $\mathbb{R}^4$ with $(z_1, z_2) = (x_1, y_1, x_2, y_2)$. We obtain

$$\begin{aligned} \dot{x}_1 &= x_1 x_2 + y_1 y_2 + x_1(\mu_1 + e_{11} r_1^2 + e_{12} r_2^2) \\ \dot{y}_1 &= x_1 y_2 - y_1 x_2 + y_1(\mu_1 + e_{11} r_1^2 + e_{12} r_2^2) \\ \dot{x}_2 &= -(x_1^2 - y_1^2) + x_2(\mu_2 + e_{21} r_1^2 + e_{22} r_2^2) \\ \dot{y}_2 &= -2x_1 y_1 + y_2(\mu_2 + e_{21} r_1^2 + e_{22} r_2^2) \end{aligned} , \tag{46}$$

where $r_j^2 = x_j^2 + y_j^2$. Linearizing about $(0, 0, \sqrt{-\mu_2/e_{22}}, 0)$, we get the diagonal matrix

$$\begin{bmatrix} \lambda_+ & 0 & 0 & 0 \\ 0 & \lambda_- & 0 & 0 \\ 0 & 0 & -2\mu_2 & 0 \\ 0 & 0 & 0 & 0 \end{bmatrix} , \tag{47}$$

where

$$\lambda_\pm = \mu_1 - \frac{\mu_2 e_{12}}{e_{22}} \pm \sqrt{-\frac{\mu_2}{e_{22}}}. \tag{48}$$

Because these equations are equivariant under rotation about the origin, this structure will be preserved around the circle of equilibria. In light of our assumptions on the parameters, we have a two dimensional stable and a one dimensional unstable manifold emanating from each point of the circle. Furthermore by looking at equations (46), one sees that the real subspace, *span*$\{ (1, 0, 0, 0), (0, 0, 1, 0) \}$, is invariant. For the moment, we focus on the

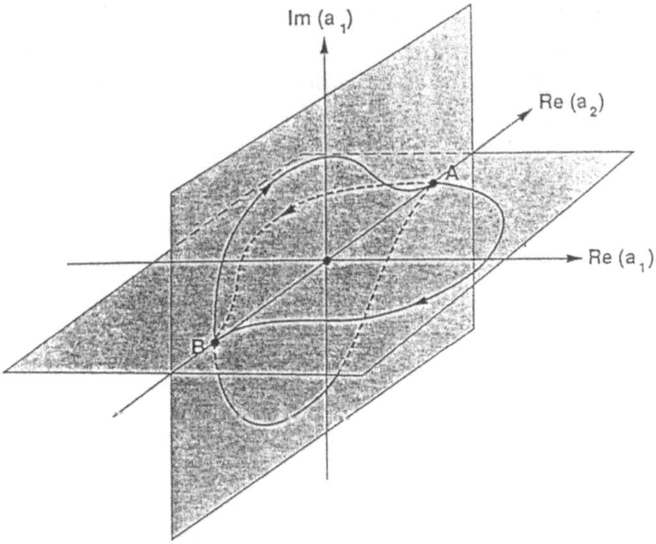

Figure 8. A and B are connected by a pair of symmetric heteroclinic cycles.

two points where the circle of fixed points intersects this invariant plane, namely $A = (0, 0, \sqrt{-\frac{\mu_2}{e_{22}}}, 0)$ and $B = (0, 0, -\sqrt{-\frac{\mu_2}{e_{22}}}, 0)$. Looking at the equations as expressed in (46) and setting $y_1 = y_2 = 0$, we see that A is a saddle and B a sink, and we find orbits leaving A are trapped in a bounded disc and that since there are no equilibria or limit sets in the half spaces $\{x_1 > 0\}$ and $\{x_1 < 0\}$, they must limit on B. Thus we have a heteroclinic orbit from A to B. By judicious application of the $O(2)$ symmetry given by transformation (38), we can complete the picture. The application of $T_{\pi/2}$ amounts to a rotation by $\pi/2$ in the z_1 plane and by π in the z_2 plane. In coordinates, this gives

$$T_{\pi/2} : (x_1, y_1, x_2, y_2) \mapsto (y_1, -x_1, -x_2, -y_2) \; .$$

Application of $T_{\pi/2}$ permutes the connections from A to B, giving a connection from B to A living in the plane defined by $\{y_2 = x_1 = 0\}$. Thus we establish the presence of a heteroclinic cycle containing A and B (see figure 8). Application of T_α for arbitrary α shows that every diametrically opposed pair of equilibria on the circle $|z_2| = \sqrt{-\mu_2/e_{22}}$ is connected in such a heteroclinic cycle.

Determining the asymptotic stability of this connection is a subtle and involved process and we refer the reader to Armbruster *et al.* (1988) for the details. It turns out that the stability is determined solely by the ratio of λ_-/λ_+. If $-\lambda_-/\lambda_+ > 1$, the heteroclinic cycle is attracting while if $-\lambda_-/\lambda_+ < 1$ it is unstable. $\square$

It should be emphasized that the heteroclinic cycles exist for an open set of parameter values. Any perturbations which respect the $O(2)$ symmetry will leave the equations, up to third order, in the functional form dictated by (45). Thus, a small perturbation, measured in function space, will only give rise to a small change in the coefficients and therefore generically will leave the heteroclinic cycle intact.

As noted in section 7.3, a similar heteroclinic cycle occurs in the 5-mode model of Aubry *et al.* (1988), and is responsible for the solutions shown in figure 5. Here the circle of equilibria lies in the (a_2, a_4) even mode subspace, intersecting the real subspace at the points $(x_1, x_2, x_3, x_4, x_5) = (0, \mp\rho_2, 0, -\rho_4, 0)$, and the unstable manifolds of each point are two dimensional. Within the real subspace, A is a saddle and B a sink; the unstable manifold of A lies in the (x_1, x_3, x_5) subspace and a connection from A to B can easily be observed numerically, although a rigorous construction involving positively invariant regions is still lacking. Accepting this "first" connection, however, the cycle follows just as with the simpler four dimensional case above.

With the above structures in hand, the model's "bursting" and "sweeping" behavior is understandable in terms of motions in the phase plane. As described in section (7.3), bursting and sweeping correspond to the rapid transitions from the neighborhood of one fixed point to that of another along a trajectory shadowing a heteroclinic connection. In the light of the symmetries of the problem and the above discussion, such connections are quite natural and generic. The even modes experience a net phase shift after each circuit, reminiscent of the observed spatial displacement between successive burst/sweep events. However, there is no particular time scale associated with attracting cycles: solutions spend successively longer periods lingering near the saddle points as they approach the cycle. We address this when we consider the restoration of the pressure term, in section 11.

10. The Kuramoto-Sivashinsky Equation

We now bricfly turn to an example which shares the $O(2)$ symmetry of the boundary layer model and exhibits heteroclinic cycles, but which is considerably simpler and easier to understand than the latter, namely the one dimensional Kuramoto-Sivashinsky (KS) equation,

$$N(u) = u_t + u_{xx} + u_{xxxx} + \frac{1}{2}(u_x)^2 = 0, \tag{49}$$

with L-periodic boundary conditions. Originally this equation was proposed in a model for instabilities in flame fronts (Sivashinsky, 1977) and for "phase turbulence" in chemical reactions. The KS equation and related equations have been used to model various interfacial instabilities, and a multi-dimensional version has been proposed to model weak fluid turbulence (Hyman *et al.*, 1986). We make no claim that the one dimensional KS equation has any direct relevance to turbulence; however, there are some parallels with the NS equations. The u_{xx} term is comparable to $\mathbf{u} \cdot \nabla \mathbf{U}$, as both introduce energy into the system and consequently destabilize it. Similarly, the u_{xxxx} term, like the dissipative $-\nu\nabla^2\mathbf{u}$ NS term, removes energy from the system and provides a stabilizing effect. In contrast to the NS equations, however, in the KS system, we can complete our analytical program, justifying almost all of our steps rigorously.

10.1. GALERKIN PROJECTION

The KS equation is invariant under translations and reflections in x. As observed in the previous section, this implies that the POD eigenbasis will simply be a Fourier basis, and that the system of ODEs resulting from projecting onto the basis will be $O(2)$-equivariant. Setting

$$u(x,t) = \sum a_j(t)\phi_j(x) = \sum a_j(t)e^{ij2\pi x/L}, \tag{50}$$

we perform a Galerkin projection to obtain

$$\begin{aligned} \textstyle\int_0^L \{\sum_k[\dot{a}_k\phi_k - (\frac{2\pi k}{L})^2 a_k\phi_k + (\frac{2\pi k}{L})^4 a_k\phi_k + \\ \textstyle\frac{1}{2}(\frac{2\pi i k}{L} a_k\phi_k)\sum_j(\frac{2\pi i j}{L} a_j\phi_j)]\,\phi_l^*\}\,dx &= 0\,. \end{aligned} \tag{51}$$

The first three terms reduce by the orthogonality property of the Fourier basis:

$$(\phi_k,\phi_l) = \int_0^L e^{\frac{2\pi i(k-l)x}{L}}\,dx = \delta_{kl}L\,. \tag{52}$$

After some rearrangement, the last term is simplified by the same reasoning:

$$\int_0^L \phi_k\phi_j\phi_l^*\,dx = \int_0^L e^{\frac{2\pi i(k+j-l)x}{L}}\,dx = (\phi_{j+k},\phi_l) = \delta_{(k+j)l}L. \tag{53}$$

Thus we arrive at

$$\dot{a}_l - \left(\frac{2\pi l}{L}\right)^2 a_l + \left(\frac{2\pi l}{L}\right)^4 a_l - \frac{1}{2}\left(\frac{2\pi}{L}\right)^2 \sum_j j(l-j)a_j a_{l-j} = 0\,. \tag{54}$$

By dividing through by $(2\pi/L)^2$ and rescaling time by the same factor we can obtain the following slightly less cluttered equations for the a_j's:

$$\dot{a}_l = l^2\left(1 - \left(\frac{2\pi l}{L}\right)^2\right)a_l + \frac{1}{2}\sum_j j(l-j)a_j a_{l-j}\,. \tag{55}$$

This relation applies to any order truncation we might perform. If we keep terms to up order K, then equation (55) defines an ODE on a $2K+1$

dimensional complex space. This can be reduced to a $2K$ dimensional complex space by noting that the evolution of the zeroth mode, a_0, is determined by the other modes:

$$\dot{a_0} = -\frac{1}{2}\sum_j j^2 a_j a_{-j} = -\frac{1}{2}\sum_j j^2 a_j a_j^* = -\frac{1}{2}\sum_j j^2 |a_j|^2. \tag{56}$$

Furthermore, because u is real, implying $a_j = a^*_{-j}$, we need consider only half of the remaining coefficients. Hence, we are left with a K dimensional system of complex ODEs (or a $2K$ dimensional system if viewed as real ODEs).

Note that equation (55) is equivariant under the group $O(2)$ with the representation (38), regardless of how many modes are included. Furthermore, the structure of (55) is identical to that of the linear and quadratic terms in the particular $K_1 = 0$, $N = 1$ case of the boundary layer model in (32).

10.2. BIFURCATION AND CENTER MANIFOLD REDUCTION

If we treat L as a bifurcation parameter, a linear stability analysis of the origin $u = 0$ (ie $a_l = 0$ for all $l \in \mathbb{Z}$) reveals that there is a series of bifurcations from the origin. Specifically, the linearization

$$\dot{a}_l = l^2\left(1 - \left(\frac{2\pi l}{L}\right)^2\right)a_l \tag{57}$$

implies that a pair of eigenvalues passes through zero when $L = 2\pi n$ for $n \in \mathbb{N}$. Therefore at $L = 2\pi n$, we have a two (real) dimensional center manifold, a $2(n-1)$ (real) dimensional unstable manifold and an infinite dimensional stable manifold consisting of all of the higher order modes. The double eigenvalue is characteristic of $O(2)$-equivariance. If we are interested in the behavior of our model around one of these bifurcations points, it is reasonable to use a model including the center and unstable dimensions and to appeal to a center-unstable manifold reduction to include information about the stable directions. Armbruster *et al.* (1989), from which this discussion is adapted, performed such a reduction.

For example, for $L \approx 4\pi$ we have a two real dimensional unstable space and a two dimensional center space spanned by a_1 and a_2 respectively. Of course, looking at equation (55), we are reminded that these variables are coupled to all of the remaining a_j's. Rather than simply truncate the system we choose to use a center manifold reduction to obtain analytic approximations for "stable" coordinates in terms of the center and unstable directions, locally near $L = 4\pi$. As usual, these are expressed as power series in the center manifold coordinates. Substituting back into the equations and truncating at cubic order we obtain

$$\begin{aligned} \dot{a}_1 &= (\tfrac{3}{4} + \mu)a_1 - 2a_1^* a_2 - a_1|a_2|^2, \\ \dot{a}_2 &= 16\mu a_2 + \tfrac{1}{2}a_1^2 - a_2(\tfrac{1}{2}|a_1|^2 + \tfrac{1}{3}|a_2|^2). \end{aligned} \tag{58}$$

Notice that this is a specific example of the $O(2)$ normal form encountered previously. As explained at the onset, this was expected because of the translation invariance in physical space. However, it should be noted that we would not have obtained the cubic terms had we simply truncated the Fourier series expansion. They play a rôle (and have a form) similar to that of the "feedback" term introduced in modeling the mean flow $U(x_2, t)$ in the boundary layer model, but here their origin is mathematically clear: no physical order of magnitude arguments are required.

Analysis of the dynamics and bifurcations of reduced models such as (58), and of models including higher order effects, may be found in Armbruster *et al.* (1989).

11. Noisy Heteroclinic Cycles and Timing

Thus far, our analysis of the boundary layer model has explained how it correctly mimics the bursting and sweeping events of boundary layer and seems to capture the fact that sequential burst/sweep events tend to be spatially displaced relative to one another. However, attracting heteroclinic cycles do not possess a distinct time scale; as noted at the end of section 9, they "die out." In reality, bursting events are observed to be to be sprinkled randomly in time, with distributions having fairly fat exponential tails (Holmes & Stone, 1992). Here we will see that restoration of the pressure term $\xi(t)$ in (32), modelled by Gaussian white noise, can produce precisely this behavior. (We note that the effect of the pressure term is not sensitive to the particular white noise form of the forcing discussed here; Stone & Holmes (1989), using a time series for $\xi(t)$ obtained from numerical data of Moin, found qualitatively similar results.)

Up to this point in our analysis, we have ignored the effect of the pressure term, $\xi(t)$, arising from the free boundary conditions at the top of the boundary layer. As mentioned during the extraction of the 5 mode model, direct numerical simulations suggest that this term is typically two orders of magnitude smaller than the a_i's. For the majority of the phase space, this amounts to a negligible perturbation. However in the neighborhood of a saddle point, the a_i's are considerably smaller in magnitude than in the rest of the flow and the effect of $\xi(t)$ is no longer negligible.

Though simulations show $\xi(t)$ to be correlated on short time scales and across (spanwise) Fourier wavenumbers, as a first approximation we will take $\xi(t)$ to be standard i.i.d. white noise which is uncorrelated in time. We will now briefly discuss the behavior of trajectories in the vicinity of a saddle point when white noise is included. Before we do, however, it is instructive to note that this is the first moment we have turned to random perturbations to explain the system's behavior. In particular, the pulsing burst/sweep behavior seems intrinsic to the flow in the boundary layer and is not driven by perturbations from deeper in the flow, only perhaps triggered by them.

11.1. THE NOISY CONNECTION

We consider the flow in a sufficiently small neighborhood U_δ of a saddle point, so that the flow is well approximated by its linearization about the fixed point. Consequently in U_δ, we model the flow by

$$\begin{aligned} dx_1 &= \lambda_s x_1 dt + \varepsilon\, dW_1(t), \\ dx_2 &= \lambda_u x_2 dt + \varepsilon\, dW_2(t), \end{aligned} \tag{59}$$

where dW_i is the formal derivative of a standard Brownian motion $W_i(t)$. This is an example of the well-studied Ornstein-Uhlenbeck (OU) process (see Arnold, 1974). Outside of U_δ, we neglect the random perturbations, supposing that during the bounded time interval needed to traverse the rest of the heteroclinic connection the deterministic part of the flow dominates. To simplify the analysis further, we also ignore the finite distortion which a distribution would experience between the time it leaves U_δ along the unstable eigenvalue and when it re-enters it along the stable eigenvalue. Detailed analysis of this system can be found in Stone & Holmes (1990). Below, we only sketch the main results in the context of the boundary layer model.

Solving Kolmogorov's equation for the one dimensional OU process, one finds the equilibrium distribution, conditioned on being at x_0 at $t = 0$, to be a mean zero Gaussian distribution, namely

$$p(x,t|x = x_0) = \mathcal{N}(x_0 e^{\lambda t}, \frac{\varepsilon^2}{2\lambda}(e^{2\lambda t} - 1)) \; . \tag{60}$$

Our model is built of two uncoupled OU processes: one stable and one unstable, corresponding to $\lambda_s < 0$ and $\lambda_u > 0$ respectively. We are interested in the competition between the two effects. We begin by assuming $-\lambda_s > \lambda_u > 0$ to match the situation in the boundary layer model (attracting cycles). Analysis shows that even with the noise, $\langle x_j \rangle \to 0$ as $t \to \infty$: the stable direction still wins. But how are the passage times through U_δ affected by the noise terms? Since we neglect any distortion outside of U_δ, we need only study how a Gaussian distribution with mean zero will evolve in the unstable direction when injected to U_δ along the stable direction.

From (60), we see that the stable OU process possesses $\mathcal{N}(0, -\frac{\varepsilon^2}{2\lambda_s})$ as a limiting distribution. Thus we assume that the solution "particle packet" has this distribution when entering along the stable direction. It will then be ejected in the unstable direction and evolve according to

$$p(x, t|\mathcal{N}(0, -\frac{\varepsilon^2}{2\lambda_s})) = \mathcal{N}(0, -\frac{\varepsilon^2}{2\lambda_s}e^{2\lambda_u t} + \frac{\varepsilon^2}{2\lambda_u}(e^{2\lambda_u t} - 1)) . \qquad (61)$$

This can be seen by thinking of the initial Gaussian profile as having evolved from a delta density existing at time $t' < 0$ such that it achieves the distribution $\mathcal{N}(0, -\frac{\varepsilon^2}{2\lambda_s})$ at $t = 0$. Thus, the result of the Gaussian distribution evolved for a time t is the same as that of a delta distribution evolved for a time $\tau = t + |t'|$. More explicitly:

$$\begin{aligned} p(x, t|\mathcal{N}(0, -\frac{\varepsilon^2}{2\lambda_s})) &= p(x, t|x(t') = x(t - \tau) = 0) = p(x, \tau|x(0) = 0) \\ &= \mathcal{N}(0, \frac{\varepsilon^2}{2\lambda_u}(e^{2\lambda_u \tau} - 1)) \; ; \quad \tau = t + \frac{1}{2\lambda_u}\ln(1 - \frac{\lambda_u}{\lambda_s}) . \end{aligned}$$

The probability that the solution leaves the interval $[-\delta, \delta]$ in the unstable direction at or before time s is

$$P_{escape}(s) = 1 - \int_{-\delta}^{\delta} p(x_2, s|x_2 = 0)\, dx_2 \; ; \qquad (62)$$

from this, the mean passage time can be estimated. In the limit where $\varepsilon \ll \delta$, it can be shown that

$$Expected(T_{total}) = K_1 + \frac{1}{\lambda_u}\{\ln(\frac{1}{\varepsilon}) + K_2\}, \qquad (63)$$

where T_{total} is the overall circuit time around the cycle; see Holmes *et al.* (1996).

Here K_1 and K_2 are constants which depend on the global structure of the homoclinic connection, but are independent of the magnitude of the noise, ε. K_1 is basically the time required for global passage on the cycle (the burst/sweep) while K_2 is related to the local dynamics linearized at the saddle point. For $\varepsilon = 0$ the expected escape time is unbounded, corresponding to the attracting cycle which "dies out." Introduction of the noise term reduces the expected escape time as $\ln(\frac{1}{\varepsilon})$ at leading order, thereby providing a characteristic interburst time scale in our model. Furthermore, this analysis predicts a distribution of escape times with exponential tails.

Stone & Holmes (1989, 1990) have confirmed these asymptotics in numerical simulations.

11.2. NOISE AND THE BOUNDARY LAYER

In the boundary layer, the time a trajectory is trapped in the neighborhood of the fixed point corresponds to the time between sweep/burst events. In fact, there is considerable evidence for such exponential tails from experimental data. However, once the timescales of Aubry *et al.* (1988) are corrected as noted in section 7.1, the "raw" bursting rates obtained from (63) with suitable estimates for ε (the r.m.s. pressure field) are still substantially lower than those observed experimentally. Noting additionally that direct numerical simulations on the minimal flow unit of Jiménez & Moin (1991), similarly sized in $L_1 \times L_3$, also yields much slower cyclic behavior than that observed, we now believe that this is due to our spatially localized model's retention of only a single structure in the streamwise direction.

In reality, a succession of quasi-independent structures of typical streamwise extent L_1 is swept past a stationary sensor, at a velocity U characteristic of the wall-normal location of the sensor. Thus structures are sampled at a rate U/L_1 with the expected interburst duration being obtained by averaging over the probability of n structures passing in succession, none of which are bursting. Identifying $T_o = K_1$ in (63) with the duration of a burst/sweep event and $T_i = \frac{1}{\lambda_u}\{\ln(\frac{1}{\varepsilon}) + K_2\}$ with the expected interburst duration for a single structure, the probability of passage of n "quiet" structures is $(1-p)^n$, with $p = T_o/(T_o+T_i)$. This leads to an estimate of observed bursting rate at a fixed sensor of

$$Expected(T_{burst}) = \frac{L_1}{U}\cdot\frac{1}{p} = \frac{L_1}{U}\left[1 + \frac{1}{\lambda_u K_1}\left\{\ln\left(\frac{1}{\varepsilon}\right) + K_2\right\}\right], \tag{64}$$

in terms of the quantities in (63). Insertion of physically reasonable estimates for L_1 and U in (64) yields bursting rates within the range of experimental observation. See Podvin *et al.* (1997) or Holmes *et al.* (1996) for details and further discussion.

It must also be said that experimental and numerical data shows that the pressure is far from a delta correlated white noise process (see Stone & Holmes (1989) for remarks and analysis of numerical simulations). Nonetheless, the analysis suggests how quasi-random fluctuations might give rise to a distribution of passage times consistent with what is observed.

12. Conclusion

In the space available to us in these lecture notes, we have only been able to sketch some of the mathematical and physical elements which form the basis of a model for the dynamics of coherent structures, in particular in boundary layer turbulence. A more comprehensive and critical discussion is found in the book by Holmes *et al.* (1996). We have attempted to emphasize that the modeling process described here can yield at best a crude model, or cartoon, of the intricate processes occurring in the boundary layer, and much physics must inevitably be averaged out, modeled, lost in the low dimensional truncation or simply ignored. In order to obtain reasonably tractable, low dimensional models, we pay the price of only capturing the behavior of the most energetic coherent structures; moreover, we model only a localized spatial domain of spanwise and streamwise extent comparable to that of an individual burst/sweep event, and thus artificially "extract" a single representative from a stream of evolving and interacting coherent structures.

Nevertheless, in spite of these simplifications and approximations, the models described here *do* seem to create a basis for understanding the dynamics of such structures. Through the technique of proper orthogonal decomposition, we are able to identify the most energetic modes, onto which we project by the Galerkin method, together with modeling of neglected modes and boundary conditions, to obtain a reasonably tractable low dimensional dynamical system. Numerical simulation, combined with analysis of the symmetries of the system and of simpler, related equations, enables us to gain some understanding of the system behavior.

In particular, the presence of stable heteroclinic cycles may be established, which appears to account for the observed burst/sweep behavior in the boundary layer. Further inclusion of random noise, modelling the influence of the pressure at the outer boundary of the wall region, yields a prediction of an exponential distribution of burst times. This agrees well with what is detected experimentally. Numerical investigations of systems including substantially more modes, in a less violent truncation, give results generally consistent with these observations. In sum, the present approach to low dimensional modeling of turbulence forms the foundation for a new approach to the understanding of the dynamics of coherent structures in certain turbulent flows.

12.1. ACKNOWLEDGEMENTS

These notes were prepared by JCM and RWW from lectures given at the Newton Institute, Cambridge, UK, in August 1995 by PJH. The authors wish particularly to acknowledge J.L. Lumley and G. Berkooz, whose forthcoming book with PJH (Holmes, Lumley & Berkooz, 1996) formed a basis for the lectures and the present article. This research was variously supported from 1985 to the present by ONR, AFOSR, NSF and DoE.

References

Armbruster, D., Guckenheimer, J. & Holmes, P. (1988) Heteroclinic cycles and modulated travelling waves in systems with $O(2)$ symmetry, *Physica D* **29**, pp. 257–282.

Armbruster, D., Guckenheimer, J. & Holmes, P. (1989) Kuramoto-Sivashinsky dynamics on the center unstable manifold, *SIAM J. Applied Math.* **49**, pp. 676–691.

Arnold, L. (1974) *Stochastic Differential Equations*. John Wiley & Sons, New York.

Aubry, N., Holmes, P., Lumley, J.L. & Stone, E. (1988) The dynamics of coherent structures in a turbulent boundary layer, *J. Fluid Mech.* **192**, pp. 115–173.

Aubry, N. & Sanghi, S. (1991) Bifurcations and bursting of streaks in the turbulent wall layer. In Lesieur, M. and Métais, O. (eds.), *Turbulence and Coherent Structures*, pp. 227–251. Kluwer, Dordrecht Boston.

Berkooz, G. (1991) *Turbulence, Coherent Structures, and Low Dimensional Models*. PhD thesis, Cornell University.

Berkooz, G. (1994) An observation on probability density equations, or, When do simulations reproduce statistics? *Nonlinearity* **7**, pp. 313–328.

Berkooz, G., Holmes, P. & Lumley, J.L. (1993a) The proper orthogonal decomposition in the analysis of turbulent flows, *Annu. Rev. Fluid Mech.* **25**, pp. 539–575.

Berkooz, G., Holmes, P. & Lumley, J.L. (1993b) On the relation between low dimensional models and the dynamics of coherent structures in the turbulent wall layer, *Theoret. Comput. Fluid Dynamics* **4**, pp. 255–269.

Brown, G.L. & Roshko, A. (1974) On density effects and large structure in turbulent mixing layers, *J. Fluid Mech.* **64**, pp. 775–816.

Cantwell, B. (1990) Future directions in turbulence research and the role of organized motion. In Lumley, J.L. (ed.), *Whither Turbulence? Turbulence at the Crossroads*, pp. 97–131. Springer-Verlag, Berlin Heidelberg.

Doering, C.R. & Gibbon, J.D. (1995) *Applied Analysis of the Navier-Stokes Equations*. Cambridge University Press, Cambridge, UK.

Foias, C. & Temam, R. (1989) Gevrey class regularity for the solutions of the Navier-Stokes equations. *J. Functional Anal.* **87**, pp. 359–369.

Guckenheimer, J. & Holmes, P. (1983) *Nonlinear Oscillations, Dynamical Systems and Bifurcations of Vector Fields*. Springer-Verlag, New York.

Herzog, S. (1986) *The Large Scale Structure in the Near Wall Region of a Turbulent Pipe Flow*. PhD thesis, Cornell University.

Holmes, P. (1990) Can dynamical systems approach turbulence? In Lumley, J.L. (ed.), *Whither Turbulence? Turbulence at the Crossroads*, pp. 195–249. Springer-Verlag, Berlin Heidelberg.

Holmes, P., Lumley, J.L. & Berkooz, G. (1996) *Turbulence, Coherent Structures, Dynamical Systems and Symmetry*. Cambridge University Press, Cambridge, UK.

Holmes, P. & Stone, E. (1992) Heteroclinic cycles, exponential tails and intermittency in turbulence production. In Gatski, T.B., Sarkar, S. & Speziale, C.G. (eds.), *Studies in Turbulence*, pp. 179–189. Springer-Verlag, New York.

Hopf, E. (1948) A mathematical example displaying the features of turbulence, *Comm. Pure Appl. Math.* **1**, pp. 303–322.

Hyman, J.M., Nicolaenko, B. & Zaleski, S. (1986) Order and complexity in the Kuramoto-Sivashinsky model of weakly turbulent interfaces, *Physica D* **23**, pp. 265–292.

Jiménez J. & Moin, P. (1991) The minimal flow unit in near-wall turbulence, *J. Fluid Mech.* **225**, pp. 213–240.

Kline, S.J., Reynolds, W.C., Schraub, F.A. & Runstadler, P.W. (1967) The structure of turbulent boundary layers, *J. Fluid Mech.* **30**, pp. 741–773.

Lumley, J.L. (1967) The structure of inhomogeneous turbulence. In Yaglom, A.M. & Tatarski, V.I. (eds.), *Atmospheric Turbulence and Wave Propagation*, pp. 166–178. Nauka, Moscow.

Lumley, J.L. (ed.) (1990) *Whither Turbulence? Turbulence at the Crossroads*. Lecture Notes in Physics, vol. 357, Springer-Verlag, Berlin Heidelberg.

Moin, P. (1984) Probing turbulence via large eddy simulation, In *AIAA paper 84-0174*.

Moin, P. & Moser, R.D. (1989) Characteristic-eddy decomposition of turbulence in a channel, *J. Fluid Mech.* **200**, pp. 471–509.

Podvin, B., Gibson, J., Berkooz, G. & Lumley, J. (1997) Lagrangian and Eulerian view of the bursting period, Phys. Fluids **9**, pp. 433–437.

Robinson, S.K. (1991) Coherent motions in the turbulent boundary layer, *Ann. Rev. Fluid Mech.* **23**, pp. 601–639.

Ruelle, D. & Takens, F. (1970) On the nature of turbulence, *Comm. Math Phys.* **20**, pp. 167–192.

Sanghi, S. & Aubry, N. (1993) Mode interaction models for near-wall turbulence, *J. Fluid Mech.* **247**, pp. 455–488.

Sivashinsky, G.I. (1977) Nonlinear analysis of hydrodynamic instability in laminar flames, Part I: Derivation of the basic equations, *Acta Astronautica* **4**, pp. 1176–1206.

Stone, E. & Holmes, P. (1989) Noise induced intermittency in a model of a turbulent boundary layer, *Physica D* **37**, pp. 20–32.

Stone, E. & Holmes, P. (1990) Random perturbations of heteroclinic cycles, *SIAM J. Applied Math.* **50**, pp. 726–743.

Temam, R. (1988) *Infinite-Dimensional Dynamical Systems in Mechanics and Physics.* Springer-Verlag, New York.

Tennekes, H. & Lumley, J.L. (1972) *A First Course in Turbulence.* The MIT Press, Cambridge, MA.

<u>GPSR Compliance</u>

The European Union's (EU) General Product Safety Regulation (GPSR) is a set of rules that requires consumer products to be safe and our obligations to ensure this.

If you have any concerns about our products, you can contact us on ProductSafety@springernature.com

In case Publisher is established outside the EU, the EU authorized representative is:

Springer Nature Customer Service Center GmbH
Europaplatz 3
69115 Heidelberg, Germany

Batch number: 10614744

Printed by Printforce, the Netherlands